The Guinea Pig

This book summarizes the development and statistical validation of a guinea pig model as an alternative for potency testing of the viral antigens included in combined vaccines applied in cattle to control the respiratory, reproductive, and neonatal calf diarrhea syndromes. The model allows, in one serum sample, to test the vaccine quality for all the viral antigens included in aqueous as well as in oil-adjuvanted formulations of bovine vaccines. The methodology proposed for the control of bovine herpes virus, parainfluenza, and rotavirus was recommended by CAMEVET as guidelines for the 30 countries in the forum, including the United States.

Key features:

- Reviews combined vaccines used for cattle
- Summarizes animal models used for vaccine testing
- Focuses on bovine herpesviruses, rotaviruses, parainfluenza, and bovine viral diarrhea virus
- Provides guidance on the effectiveness of the guinea pig model for testing vaccine immunogenicity

POCKET GUIDES TO
BIOMEDICAL SCIENCES

The ***Pocket Guides to Biomedical Sciences*** series is designed to provide a concise, state-of-the-art, and authoritative coverage on topics that are of interest to undergraduate and graduate students of biomedical majors, health professionals with limited time to conduct their own searches, and the general public who are seeking for reliable, trustworthy information in biomedical fields.

The Guinea Pig Model

An Alternative Method for Vaccine Potency Testing

Viviana Parreño

CRC Press is an imprint of the
Taylor & Francis Group, an **informa** business

First edition published 2023
by CRC Press
6000 Broken Sound Parkway NW, Suite 300, Boca Raton, FL 33487-2742

and by CRC Press
4 Park Square, Milton Park, Abingdon, Oxon, OX14 4RN
CRC Press is an imprint of Taylor & Francis Group, LLC

ISBN: 9781032446400 (hbk)
ISBN: 9780367489861 (pbk)
ISBN: 9781003373148 (ebk)

DOI: 10.1201/9781003373148

Typeset Frutiger
by Deanta Global Publishing Services, Chennai, India

Contents

Series Editor Introduction

The Pocket Guides to Biomedical Sciences series is designed to provide concise, state-of-the-art, and authoritative coverage on topics that are of interest to undergraduate and graduate students of biomedical majors, health professionals with limited time to conduct their own literature searches, and the general public who are seeking reliable, trustworthy information in biomedical fields. Since its inauguration in 2017, the series has published 12 books (https://www.routledge.com/Pocket-Guides-to-Biomedical-Sciences/book-series/CRCPOCGUITOB) that cover different areas of biomedical sciences. The most recent two titles form unique sister pair volumes: *Vaccine Efficacy Evaluation: The Gnotobiotic Pig Model* and *The Guinea Pig Model: An Alternative Method for Vaccine Potency Testing*. In these two books, the authors review their decades-long research efforts in the development of two unconventional animal models for vaccine development, evaluation, and quality control. Testing the immunogenicity, protective efficacy, and safety in animal models is one of the most important steps in vaccine development after the construction and formulation of the protective antigens and before human clinical trials. The pig (*Sus scrofa*) has high similarities with humans in gastrointestinal anatomy, physiology, nutritional/dietary requirements, and mucosal immunity. For preclinical testing of human rotavirus and norovirus vaccines, an animal model that can exhibit the same or similar clinical signs of disease as humans is critical for assessing protection against both infection and disease upon challenge. Gnotobiotic pig models fulfill this need. Mouse models are more readily available than pig models and are useful for testing vaccine immunogenicity; however, mice cannot be infected by human rotavirus or human norovirus and are not useful for evaluation of vaccine-induced adaptive immunity associated with protection against human rotavirus or norovirus disease. Through the author's studies detailed in her book, the gnotobiotic pig model has been firmly established as the most reliable animal model for the preclinical evaluation of human rotavirus or human norovirus vaccines. Equally important, but for animal vaccines, the guinea pig (*Cavia porcellus*) model represents an alternative method for testing the potency of viral vaccines applied in cattle. Due to the robustness of the statistical validation, the guinea pig model has been adopted by the National Service of Animal Health of Argentina (sanitary resolution 598.12) as the official potency testing model. The recommendation and guidelines to apply the guinea pig model in the quality control of infectious bovine rhinotracheitis (caused by bovine herpesvirus type 1), rotavirus, and parainfluenza vaccines were agreed by all the member countries of the American Committee for Veterinary Medicines, focal point of the World Organisation for Animal Health in the Americas. These two books demonstrate the indispensable role of animal models in biomedical research and in developing and producing efficacious vaccines that are critical for improving human and animal health.

Lijuan Yuan

Series Editor Biography

Lijuan Yuan is a Professor of Virology and Immunology in the Department of Biomedical Sciences and Pathobiology, Virginia-Maryland College of Veterinary Medicine, Virginia Tech. Dr. Yuan studies the interactions between enteric viruses and the host immune system. Her laboratory's research interests are focused on the pathogenesis and immune responses induced by enteric viruses, especially noroviruses and rotaviruses, and on the development of safer and more effective vaccines as well as passive immune prophylaxis and therapeutics against viral gastroenteritis. These studies utilize wild-type, gene knock-out, and human gut microbiota transplanted gnotobiotic pig models of human rotavirus and norovirus infection and diseases. Dr. Yuan's research achievements over the past 29 years have established her as one among the few experts in the world leveraging the gnotobiotic pig model to study human enteric viruses and vaccines. She is the author of *Vaccine Efficacy Evaluation: The Gnotobiotic Pig Model* in this *Pocket Guides to Biomedical Sciences* series. Currently, the Yuan laboratory is evaluating the immunogenicity and protective efficacy of several candidate novel rotavirus and norovirus vaccines and engineered probiotic yeast *Saccharomyces boulardii* secreting multi-specific single-domain antibodies as novel prophylaxis against both noroviruses and *Clostridioides difficile* infection. See: https://vetmed.vt.edu/people/faculty/yuan-lijuan.html

Foreword

When Dr. Viviana Parreño asked me to write this preface, I felt honored as well as pleased due to our long-lasting collaborative work, summarized in this book. Quite a few field and laboratory trials carried out from 2003 onward were needed before methods and experimental protocols were set up.

Once experimental methods were stated by the statistician team, veterinary practitioners and farm workers got involved in the detection of breeding or dairy farms suitable for conducting the field trials in cattle before carrying out vaccinations and sampling. On the other hand, researchers and technicians – at National Institute of Agricultural Technology laboratories – helped not only with vaccination and sampling of guinea pigs and serological testing of bovine and guinea pig sera but also in the development and validation of enzyme-linked immunosorbent assays (ELISA).

As the coordinator of bovine trials and the link between a private veterinary biologics company and the Virology Institute of the National Institute of Agricultural Technology at Buenos Aires, I was pleased to work along with veterinary practitioners doing the best in cattle vaccination and sampling. At the same time, National Institute of Agricultural Technology researchers opened my mind to applied science.

Behind the figures, graphics, and tables that have been published in several scientific papers and are summarized in this book, there are hundreds of calves, cows, and heifers. And guinea pigs, of course.

Finally, as a member of Dr. Parreño's Guinea Pig Model Team, I felt proud when the first milestone showed up after years of hard work since the National Service of Animal Health of Argentina adopted the model as the official potency testing of viral veterinary vaccines, in 2012, and also when, years later, the 20 country members of Committee for Veterinary Medicines-World Organisation for Animal Health approved this model.

María Marta Vena
DVM

Acknowledgments

Statisticians
Lic. Laura Marangunich
Ing. Maria Virginia Lopez

Business specialist
Lic. Paula Garnero

Funding sources
PICT Start up0667. Agencia de promoción científica y técnica, MincyT, Argentina
https://www.argentina.gob.ar/ciencia/agencia/la-agencia

Collaborators
National Institute of Agricultural Technology, INTA, Argentina
https://www.argentina.gob.ar/inta
Dr. Fernando Fernández
Dr. Marina Bok
Dr. Celina Vega
Dr. Alejandra Romera
Dr. Silvina Maidana
Dr. Marina Mozgovoj
Dr. Andrea Pecora
Dr. Darío Malacari
Dr. María Sol Aguirreburrualde
Dr. Anselmo Odeón
Dr. Gisela Marcoppido
Ing. Lucia Rocha
Lic. Josefina Baztarrica
Tec. Diego Franco
Jose Vallejos
Daniela Rodriguez
Nancy Suarez Perez

National Animal Health Authority of Argentina (SENASA)
https://www.argentina.gob.ar/senasa
Dr. Eduardo Maradei
Dr. Virginia Barros
DVM Valeria González Thomas
DVM Pilar Muntadas

Veterinary vaccine companies

Contributing to the development of the calibrator vaccines used in the statistical validation of the model

Biogénesis Bagó S.A.

Dr. María Marta Vena
Dr. Mercedes Izuel
DVM Jorge Fillippi

CDV S.A.

Dr. Alejandro Menyhard
Jose Runcio
Dr. Pablo Halperin

Vetanco S.A.

Dr. Demian Bellido
Ing. Jorge Winokur

PROSAIA Foundation, Argentina

www.prosaia.org
Dr. Javier Pardo
Dr. Alejandro Schudel
Dr. Carlos Van Gelderen

Ad hoc group of biologists

Collaborators in the writing of PROSAIA IBR, PI-3, and RVA Guidelines submitted to CAMEVET

Dr. María Marta Vena (Freelance consultant)
Dr. Eduardo Mortola (La Plata University)

Cámara Argentina de la Industria de Productos Veterinarios (CAPROVE)

Dr. Enrique Argento
Dr. Alejandro Ham
Dr. Marianna Loppolo
Dr. Walter Ostermann
Dr. Eliana Smitsaart

Cámara de Laboratorios Argentinos Medicinales Veterinarios (CLAMEVET)

Dr. Hugo Gleser

Collaborators in the writing of PROSAIA BVDV Guideline submitted to CAMEVET

DVM Gustavo Combessies (Laboratorio Azul S.A.)

DVM Fernando Fernández (National coordinator of research and development, INTA)
DVM Valeria González Thomas (Laboratory animal director of National Agri-Food Quality and Health Service – SENASA)
DVM Alejandro Ham (Biogénesis Bagó S.A. – CAPROVE)
Dr. Darío Malacari (Laboratorio CRESAL Veterinaria S.A.)
Dr. Anselmo Odeón (EEA INTA Balcarce)
Dr. Andrea Pecora (Instituto de Virología, CICVyA, INTA. Assistant researcher, CONICET)
Dr. María Sol Perez Aguirreburualde (Instituto Patobiología, CICVyA, INTA)
Dr. Sandra Tobolski (Biogénesis Bagó S.A. – CAPROVE)
DVM María Marta Vena (independent consultant)
Dr. Andres Wigdorovitz (INCUINTA. CICVyA, INTA Castelar. Principal researcher, CONICET)
Javier Pardo. DVM, PROSAIA Foundation) was in charge of the coordination of the ad hoc groups in all cases

Countries participating in the guidelines' writing groups at CAMEVET

IBR, PI-3, and RVA Guidelines

Argentina (Virginia Barros, and Ricardo D'Aloia, SENASA)
Brazil (Marcos Vinicius)
Chile (Dr. Fernando Zambrano)
Colombia (Nestor Guerrero Lozano)
United States (Rick Hill)

BVDV Guideline

Argentina (INTA) (Dra. Viviana Parreño, Dra. Andrea Pecora, INTA)
Argentina (Carlos Francia, CAPROVE)
Brazil (Dr. Emilio Salani, SINDAN)
Chile (Dr. Fernando Zambrano)
Colombia (Dr. Nestor Guerrero Lozano)
Mexico (Dra. Ofelia Flores)
Mexico (Lic. Rocio Alexandra Luna Orta, INFARVET)
Uruguay (Dra. Mercedes Echeverry, CEV)

I am especially grateful to Dr. María Marta Vena, my colleague and friend, who always encouraged me and worked very hard beside me to conduct this project during all these years to take it to PROSAIA and CAMEVET. Finally, I also thank Dr. Lijuan Yuan, my mentor and friend, who encouraged and supported me during my visit to Virginia Tech to write this book.

About the Author

Viviana Parreño, Biochemist (1995), PhD in Veterinary Virology and Immunology (2002) and Master in Biometrics (2021), has worked since 1992 at The National Institute of Agricultural Technology, Argentina, and has been a member of the National Scientific and Technical Research Council in her scientific career since 2007. At the moment of writing the present book, Dr. Parreño had been working as a visiting research scientist at the Department of Biomedical Sciences and Pathobiology, Virginia-Maryland College of Veterinary Medicine, Virginia Polytechnic Institute and State University from July 2021 to October 2022.

Dr Parreño was in charge of the VD-Lab (viral gastroenteritis lab) at the Virology Institute, National Institute of Agricultural Technology, from 2004 to 2016. She has 30 years of research experience, including epidemiological studies of animal rotavirus, coronavirus, and norovirus; development of vaccines and passive immune strategies to prevent calf and foal diarrhea; development of calf models of bovine rotavirus and coronavirus infection and disease; development and characterization of llama-derived nanobodies; and development and statistical validation of diagnostic kits, including Rotadial, the first nanobody-based kit used in the hospitals of the acute diarrhea surveillance network of the Argentinean Ministry of Health (INEI-ANLIS Malbran).

Dr. Parreño is co-founder and scientific coordinator of INCUINTA (2009 to present), the incubator of biotechnological platforms of the National Institute of Agricultural Technology. She is one of the entrepreneurs who created the public–private startup company BIOINNOVO (INTA-Vetanco S.A.) and participated actively in the development of three veterinary products on the market: Rotacoli Equina® Biochemic; BIOINNOVO IgY DNT®, an IgY-based product to prevent calf diarrhea due to rotavirus A, coronavirus, ***Escherichia coli***, and ***Salmonell***a; and VEDEVAX®, a bovine viral diarrhea virus subunit targeted vaccine.

From 2003 to the present, Dr. Parreño has coordinated a large team of researchers and private partners to achieve the development and statistical validation of the National Institute of Agricultural Technology (INTA) guinea pig model for potency testing of bovine viral vaccines, the topic of this present book. The model was adopted as the official potency control test of vaccines administered to cattle in Argentina (National Animal Health Authority of Argentina – Animal Health Service Resol. 598.12), and its use is expected to be extended through the American Committee for Veterinary Medicines (CAMEVET) and the World Organisation for Animal Health at a regional and worldwide level.

1
Vaccination of Cattle

1.1 Vaccines and vaccination programs

Vaccination programs for beef and dairy cattle are designed to protect the animals from diseases caused by pathogens, including viruses, bacteria, and protozoa. Vaccines are administered to the animal in order to stimulate its immune system to produce a protective response against several antigens at the same time. We call these immunogens combined or multivalent vaccines. The immune system will then "remember" how to produce a response against each pathogen if it ever is infected with any of them. Vaccines teach the immune system and help animals to be prepared to fight against an infection. Some vaccines can provide complete protection (sterile immunity), like the ones against food and mouth disease (FMD) (Rawdon et al. 2018). Other vaccines will allow a subclinical infection or the development of a less severe disease (Dubovi et al. 2000).

1.2 Vaccine types used in cattle

Vaccines designed for cattle include modified live vaccines (MLVs) or killed/inactivated vaccines.

MLVs contain a small amount of virus (Chamorro et al. 2016) or bacteria (Wang et al. 2022) that has been modified so that it does not cause clinical disease when used according to product label instructions. However, the virus or bacteria can still replicate in the vaccinated animal, resulting in a controlled attenuated infection (Walz et al. 2017). Recognition of the replicating organism by the animal's immune system stimulates an effective immune response including B-cell and T-cell responses. MLVs are mainly available for diseases caused by viruses, such as infectious bovine rhinotracheitis (IBR), bovine viral diarrhea virus (BVDV), parainfluenza-3 (PI-3), and bovine respiratory syncytial virus (BRSV). Some MLVs are safe for use in pregnant cows if you follow the recommendation of the manufacturer. However, if not used correctly, MLVs for BVDV can produce abortion in pregnant cows, for example. Some MLVs are not approved for use in calves nursing pregnant cows because of the slight possibility that the calves could temporarily shed the vaccine virus and infect the cows. However,

DOI: 10.1201/9781003373148-1

some MLVs can be safely used in calves nursing pregnant cows if the cows have been vaccinated following label directions. MLVs are also safe to use in weaned calves, including heifers. The great advantage of MLVs is that they induce B-cell and T-cell responses with the development of effective antibody and cytotoxic responses (Chamorro and Palomares 2020; Griebel 2015; Reppert et al. 2019).

Killed (inactivated) vaccines usually contain chemically inactivated pathogens, such as an inactivated virus or bacteria (bacterin). They can also contain toxoids, inactivated toxins. Currently, there also exist subunit vaccines, which are usually composed of a highly immunogenic protein of the pathogen expressed as a recombinant protein (Bellido et al. 2021). These killed/subunit vaccines do not replicate in the animal after administration, so they are highly safe because no reversion to virulence is possible. However, they usually contain adjuvants, chemicals that cause inflammation in the inoculation site and stimulate the immune system to induce a proper immune response. On the other hand, inactivated vaccines used in cattle have the limitation that they only induce antibody (Ab) responses but not active cytotoxic T-cell responses. The immunity induced by these types of vaccines is not long-lasting and requires booster doses to maintain the Ab titers at protective levels. Finally, their great advantage is their safety; they can be used in any animal category, including pregnant cows (Dubovi et al. 2000; Patel 2004).

1.3 DIVA vaccines

DIVA stands for Differentiating Infected from Vaccinated Animals. For example, this can be achieved by using a vaccine based on a virus strain with the deletion of a viral protein that is not indispensable for virus replication. An example of this is glycoprotein E (gE) from bovine herpesvirus 1 (BoHV-1), the etiological agent of bovine rhinotracheitis (IBR). This gE-virus as vaccine antigen can be used together with a serological test that can differentiate animals with vaccine-induced antibodies (e.g. against the neutralizing antigens but negative for gE) from infected animals with antibodies against the field virus (gE positive) (Romera et al. 2014). This type of vaccine has been crucial in the success of eradication campaigns in Europe (Muratore et al. 2017).

1.4 Combined vaccines for cattle

Most of the vaccines used in cattle are combined vaccines that have been designed for the prevention and control of cattle diseases, which can be part of the respiratory, reproductive, and neonatal calf diarrhea syndromes. A large number of combined vaccines are available in the market. These multivalent formulations have been used for many decades in routine vaccination protocols to prevent bovine respiratory, reproductive, or diarrheic syndromes in cattle. They were designed to facilitate the application of the immunization calendar and to control a sanitary problem of complex etiology, simultaneously.

Combined vaccines applied in cattle mostly include inactivated viruses; however, there are some options mixing inactivated with live attenuated viruses (for example, the vaccine Cattle Master from Zoetis Animal Health) or monovalent targeted subunit vaccines (E2 protein of BVDV in the VEDEVAX vaccine, Bioinnovo S.A.) (Bellido et al. 2021). Parenteral administration is the most frequent administration route, although some vaccines may be given by other routes, such as intranasal and oral (i.e. Inforce3 and Calf Guard, Zoetis).

Multivalent vaccines usually include in their complex formulations aqueous or oil adjuvants combined with bacterins (for example, some of the following: *Pasteurella multocida*, *Mannheimia haemolytica*, *Histophilus sommi*, *Moraxella bovis*, *Branhamella ovis*, *Campylobacter fetus-fetus*, *Campylobacter fetus-venerealis*, and *Leptospira* sp.) and some or all of the following viral antigens: IBR, BVDV for reproductive vaccines; IBR, BVDV, parainfluenza virus type 3 (PI-3), BRSV for respiratory vaccines. The vaccines designed to prevent neonatal calf diarrhea usually contain *Escherichia coli* with virulent factors, *Salmonella dublin* and *S. typhimurium* together with bovine rotavirus and coronavirus (Figure 1.1).

This book will be focused on testing these combined multivalent vaccines regarding their potency for the viral antigens included in their formulation.

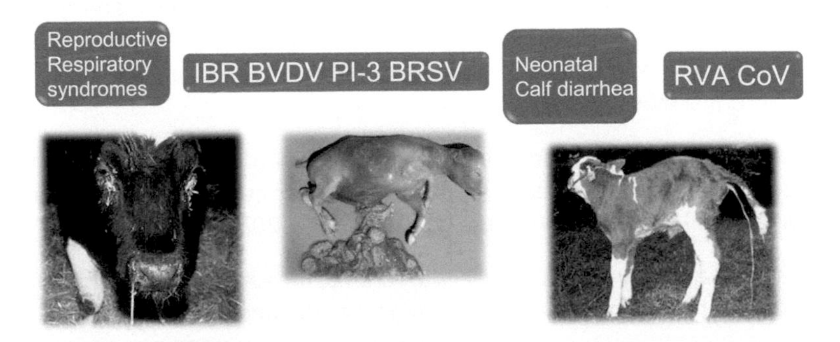

Figure 1.1 Viral agents associated with the reproductive, respiratory, and neonatal calf diarrhea syndromes that are included in the combined vaccines applied in cattle.

1.5 Reproductive and respiratory vaccines

Next, the different viral agents included in the combined vaccines for cattle will be briefly described.

BoHV-1 and BoHV-5 (bovine herpes virus 1 and 5) are both implicated in IBR. BoHV-1 is considered the causative agent of IBR, a disease resulting in respiratory signs, reproductive failure, and abortions. IBR disease is sometimes called "red nose" and often initiates the shipping fever complex. BoHV-5 is also related to neurological forms of the disease. These viruses are present in reproductive as well as in respiratory vaccines (9CFR 113.216 2014; OIE 2021).

BVDV (bovine viral diarrhea virus) causes disease resulting in numerous problems, such as damage to the digestive and immune systems, pneumonia, abortions, calf deformities, and the generation of persistently infected animals. Reproductive as well as respiratory vaccines for cattle usually carry the BVDV 1a variant. Some vaccines also include the BVDV 2 variant (9CFR 113.215 2014; OIE 2018).

PI3 (parainfluenza-3 virus) is a virus associated with respiratory disease, always included in respiratory vaccines (Makoschey and Berge 2021; 9CFR 114.309 2014).

BRSV (bovine respiratory syncytial virus) is a virus that can cause severe, acute respiratory disease, especially in young cattle. Some respiratory vaccines include BRSV (Makoschey and Berge 2021) in their formulations.

1.6 Neonatal calf diarrhea syndrome

Rotavirus A (RVA) is a virus that can cause diarrhea (scours) and dehydration in young calves. Most scours vaccines given to pregnant females include rotavirus (Blanchard 2012; Brunauer, Roch, and Conrady 2021) in their formulations.

Coronavirus is also a virus that can cause severe diarrhea, acidosis, and dehydration in young calves. Some scours vaccines given to pregnant females will contain coronavirus (Zhu, Li, and Sun 2022).

In the case of neonatal calf diarrhea syndrome, the vaccines are given to the pregnant dams in the last stage of pregnancy (60 and 30 days before calving) in order to raise the level of antibodies to rotavirus and coronavirus in their bloodstream just in time with colostrum production. An active transfer of bovine IgG1 Ab from serum to colostrum will concentrate around 10 times the IgG1 Ab titer in the dam's serum into the colostrum. Passive maternal Ab acquired via colostrum intake during the first hours of life will provide passive immunity to the newborn calf to protect it against severe diarrhea for the first few weeks or months of life (Parreño et al. 2004; Saif and Fernandez 1996).

Until 2013, there was no harmonized methodology that would allow batch-to-batch quality control of the immunogenicity and efficacy of these vaccines for each of the viral agents included therein. This could compromise the vaccine performance in the field, causing a significant detriment to the productivity of the livestock industry, especially in developing countries.

In this pocket guide, we describe a **guinea pig model** for vaccine potency testing developed at the National Institute of Agricultural Technology (INTA), Argentina. This laboratory animal model is a low-cost, fast test (potency result within 35 days, whereas the test in cattle will require 65 days at minimum) that predicts vaccine efficacy in cattle. It has been statistically validated for each virus antigen against the target species (bovine). Currently, the **INTA guinea pig model** is the official quality control test to evaluate the vaccine potency for IBR and RVA adopted by the Argentinean National Service of Animal Health (SENASA) in 2013 (sanitary resolution 598.12), and since its implementation, it has promoted a significant improvement in the quality of the vaccines produced in Argentina (see Chapter 6). Recommendation guidelines for the potency testing of vaccines for IBR, rotavirus, PI-3, and BVDV were elaborated by the ad hoc "biologic group" of specialists at PROSAIA Foundation, Argentina (Schudel, Van Gelderen, and Pardo). These potency testing protocols were submitted to the American Committee for Veterinary Medicines (CAMEVET), the focal point of the World Organisation for Animal Health (WOAH, previously OIE) in the Americas. Editorial committees formed by representatives of the official and private sectors of different countries analyzed the proposals. Finally, the protocols were approved as recommendation guidelines by all the CAMEVET member countries (CAMEVET).

The purpose of this mini-book is to review the validation of the model for four out of the six viruses present in the combined vaccines applied in cattle in order to promote its implementation by the animal health services of other countries, given that it represents an excellent tool for batch-to-batch potency testing of bovine combined vaccines.

1.7 Success of the vaccination program: not only vaccine quality counts

Before entering fully into the INTA guinea pig model validation and application, it will be important to address other critical factors, besides the vaccine quality, that must be taken into account at the moment of programming any vaccination schedule in bovine herds.

1.7.1 Proper vaccine timing

Timing and proper administration of vaccines are key factors to be considered when trying to protect a herd from disease. Considering young animals being vaccinated for the first time, a second dose is often required a few weeks after the first or primary vaccination. *An initial two-dose vaccination is definitely required for killed vaccines to provide optimal priming of the immune system.* The vaccine producer's directions will indicate if and when a second dose is required, as well as the subsequent boosters, which are usually recommended to be given annually. The time between the primary and second vaccination doses is also very important; failure to give the second dose at the proper time might result in an unprotected adult animal even if that animal is boosted every year thereafter. Management considerations might make it difficult for some farmers to give the second dose of vaccine within the time span called for on the label, which is often from 3 to 6 weeks after primary vaccination.

1.7.2 Proper vaccine conservation and handling

Proper conservation and handling of vaccines is very important. The best vaccine program will fail if the vaccine is damaged by improper handling. Cold chain is very important to warrant vaccine quality. This not only refers to the fact that most vaccines need to be stored at 4°C. Vaccines also *should not be frozen or left in direct sunlight*. Most MLV vaccines, usually lyophilized, must be reconstituted by adding sterile water into the dehydrated material provided in a separate sterile vial. After the water is added, the vaccine organisms are fragile and will be "alive" for only a short period of time. As a rule of thumb, only reconstitute enough vaccine to be used in 30–45 minutes, and use a cooler or other climate-controlled storage container to protect reconstituted vaccines from extremes of cold, heat, and sunlight (Figure 1.2).

1.7.3 Method of vaccine administration/vaccination route

The method of injection is another important issue. The only acceptable site for injection in cattle is the neck for both intramuscular (IM; in the

Figure 1.2 Always bring the vaccines in a foam box with ice packs.

muscle) and subcutaneous (SC; under the skin) routes (Figure 1.3). IM injections of some products can cause significant muscle damage, so it is necessary to avoid injecting anything into the top butt or rump of the animal causing damage to valuable beef products. Muscle damage costs the beef industry millions of dollars each year. Thus, SC is the preferred administration route for vaccines that allow IM or SC administration. Use 18- or 16-gauge needles ½ or ¾ inches long to administer SC injections, and use 18- or 16-gauge needles 1 to 1½ inches long to administer IM injections.

Keep needles and syringes clean to avoid infections at the site of injection. DO NOT use disinfectants to clean needles and syringes used to administer

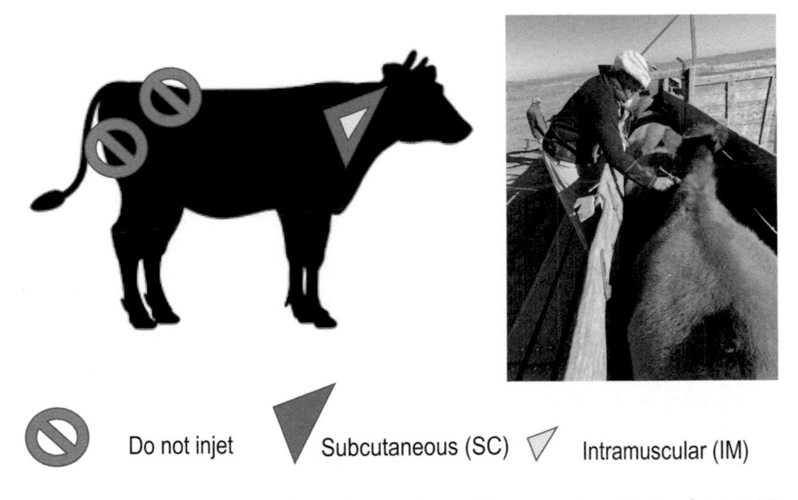

Figure 1.3 Proper injection site and route for cattle vaccination. (Taken from ANR-1280. "Alabama Beef Quality Assurance: Administer Drugs Properly".)

vaccines, especially MLVs. Even a trace or film of disinfectant in a syringe or needle can kill the live organisms and make vaccination worthless. Follow product guidelines for cleaning multi-use vaccine syringe guns, but in general, after use, rinse the syringe gun thoroughly with hot water to clean it after use, and then sterilize it using boiling water.

Finally, it is very important not to mix different vaccines into one syringe or combine other injectable drugs into the same syringe with vaccines. Although this method has been advocated as a method of reducing the number of injections, it could inactivate the vaccine because of incompatibilities with other drugs.

1.7.4 Vaccinating the right animal at the right time

Vaccination schedules are not fixed for all herds and husbandry systems, but in general, it is recommended to administer a respiratory and reproductive vaccine at weaning (two doses at a 30-day interval), a booster at rearing, and another booster at feedlot entry.

To protect heifers and cows against reproductive diseases, it is often recommended to vaccinate at least 6 to 8 weeks prior to the breeding season to allow the development of a protective immune response at the moment of male service or artificial insemination.

In the case of protecting cows' offspring from neonatal diseases such as calf scour or respiratory diseases, vaccines are given to the dam to increase the antibody titers in the colostrum. The first vaccine dose is recommended to be administered at 60 days pre-calving and the second dose around 30 days pre-calving.

Vaccine timing varies from product to product; consequently, always follow vaccine label directions with respect to vaccine administration timing to maximize product efficacy. Most recommended vaccines are best given at specific ages and/or at specific times as related to management and reproductive cycles.

1.8 Guidelines for quality evaluation of viral vaccines applied in cattle

Vaccination is a cost-effective strategy to prevent infectious diseases in cattle. Vaccines are biologicals that are used to immunize large groups of healthy animals, and they may be subject to inherent batch-to-batch variation (EMEA/140/97 1997; EMEA/P038/97 1998). Veterinary vaccine manufacturers make a great effort to obtain approval for each vaccine that they produce from the animal health authorities in each country where the vaccine is marketed. Vaccine companies must prove that the vaccine follows certain specifications, including purity (i.e. freedom from extraneous matter), potency (i.e. the capacity of a vaccine batch to exert its effect), safety (i.e. relative freedom from harmful effects), and efficacy (i.e. effect of vaccination on the target species/population under ideal circumstances) (Geletu, Usmael, and Bari 2021).

The vaccine's label will always specify the diseases and pathogens against which the vaccine provides protection. Harmonized tools are needed to ensure that the vaccine will do what its label claims (Jungbäck 2011).

Vaccines manufacturers who apply for obtaining a license (registration) for the production of veterinary vaccines containing IBR, BVDV, or PI-3 in their formulation in the United States or the European Union must demonstrate compliance with standards regulated by animal health inspection services (Animal and Plant Health Inspection Service [APHIS]; North American Code of Federal Regulations [CFR]; the European Pharmacopoeia; the European Medicines Agency [EMEA]; the Manual of Diagnostic Tests and Vaccines for Terrestrial Animals, World Organisation for Animal Health [OIE]). Regulations concerning the efficacy and immunogenicity of those vaccines require animal trials to be performed on the target species, which usually consist of the vaccination of susceptible seronegative cattle in order to evaluate the serological antibody response. Following those trials, if the vaccine-induced Ab titers are low, the animals need to be submitted to a virus challenge test to assess protection against infection (Figure 1.4). Clear recommendations can be found in the American CFR and the Manual of Diagnostic Tests and Vaccines for Terrestrial Animals of the OIE for killed IBR and BVDV vaccines (9CFR 113.215 2014; 9CFR 113.216 2014; 9CFR 114.309 2014; OIE 2022).

Once the product is approved by the official regulatory agency, the quality control of each batch to be released into the market must be performed by means of a potency test that determines the immunogenicity in cattle or in another laboratory animal model (*in vivo*) or by using *in vitro* tests validated against the vaccine testing in the target species (Figure 1.4) (EMEA/140/97 1997; EMEA/P038/97 1998). It is important to highlight that no reference regulations are as yet available to specifically evaluate the immunogenicity and efficacy of killed vaccines containing PI-3, rotavirus, coronavirus, and BRSV. Finally, vaccines must demonstrate optimal performance under field conditions. This last control is usually conducted by the manufacturer in collaboration with research institutions as a promotion of their products (Figure 1.4).

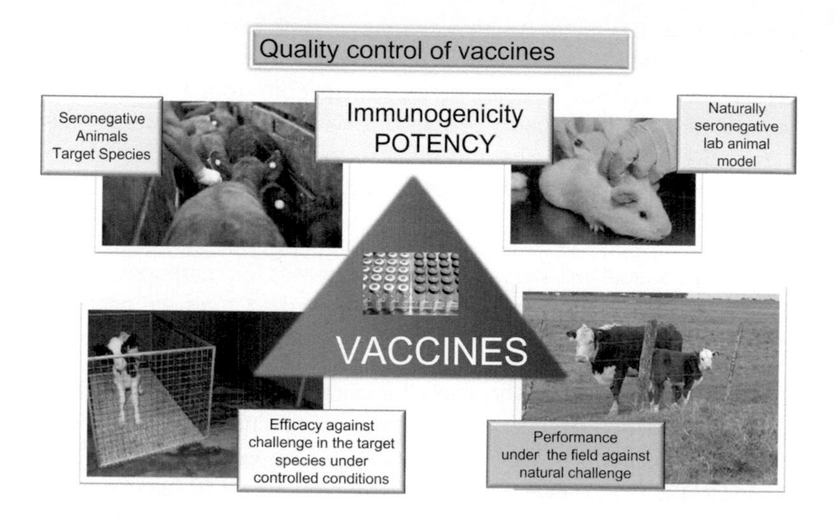

Figure 1.4 Vaccine quality control. Potency test to evaluate the immunogenicity (antibody responses) in the target species and in a well-standardized lab animal model. Efficacy can be assessed in the target species, where animals are immunized and then experimentally challenged with the pathogen to evaluate protection against infection and disease under controlled conditions. Finally, vaccine performance needs to be demonstrated under field conditions against natural challenges. Farmers must see the improvement in the health of the herd and also in the productive indexes.

1.8.1 Consistency approach and good manufacturing practices

Regarding animal welfare, international organizations encourage the development of *in vitro* tests to avoid and reduce to the minimum the use of animals for experimental control tests. Many potency tests have already gone through a transition from painful challenge tests *in vivo* to reduced and refined tests *in vivo* (i.e. fewer animals and less discomfort) to *in vitro* batch testing (Akkermans et al. 2020).

In the particular case of combined, inactivated, and adjuvanted vaccines for cattle, the application of these *in vitro* techniques is possible, and it is worth exploring them. However, given the disadvantage that each formulation (group of inactivated and live attenuated antigens, adjuvant, bacterins, toxoids, subunits, and DNA vaccines) is unique and in most cases intellectually protected by the manufacturer, the *in vitro* test will need to be customized. Then, a reduced and refined *in vivo* test is still considered inevitable to assess the potency of these products, especially for the viral antigen included, for which no recommendation is yet available to measure the potency (Akkermans et al. 2020; Jungbäck 2011; Taffs 2001).

Another important issue to be considered at the time of conducting a potency and efficacy test (i.e. the trial needed for the registration of a new vaccine) is the difficulty of finding seronegative bovines for most of the viral agents to be included in these combined vaccines, since most of these viral diseases are endemic almost worldwide. The high cost of testing for immunogenicity in the natural host does not allow potency and efficacy tests to be carried out routinely in the target species. This posed the need to develop a standardized test in a laboratory animal model that would allow the batch-to-batch potency evaluation of all kinds of vaccines in a harmonized way.

1.9 Vaccine potency testing in guinea pigs

In 2003, a group of virologists from the Virology Institute of INTA, Argentina, started the development and standardization of a test in laboratory animals (guinea pigs) as quality control of the immunogenicity for each of the viral antigens included in the combined vaccines applied in cattle. The aim was to assess the potency of each vaccine batch, guaranteeing high-quality products in the local market.

The INTA guinea pig model has been validated and its concordance evaluated for vaccine classification in comparison to the target species for four out of the six viruses included in the combined vaccines for cattle: IBR, bovine rotavirus, PI-3, and BVDV, while the validation for bovine coronavirus is still in progress (data not shown). The validation of the model also included the development and statistical validation of the corresponding enzyme-linked immunosorbent assay (ELISA) techniques to measure the bovine and guinea pig Ab responses in serum samples in comparison with the gold standard techniques according to the OIE and CFR (in the case of IBR and BVDV, virus neutralization and in the case of PI-3, hemagglutination inhibition assay [HIA]). We also developed and statistically validated the model and the associated ELISAs to evaluate the potency of vaccines for rotavirus and coronavirus, two viruses that do not have recommendations in the CFR or OIE guidelines. This mini-book is a summary of the results obtained in the validation of the guinea pig model for each virus and the results of its implementation by SENASA as the official quality control of veterinary vaccines in Argentina.

1.9.1 General immunization protocol and sampling of guinea pigs and bovines

The assay for testing viral vaccines in guinea pigs is based on the immunization of six guinea pigs (*Cavia porcellus*), SiS Al, around 350–400 g in weight, with two doses of vaccine (21 days apart). The vaccine is applied subcutaneously. In the initial studies, the animals were immunized using different dose volumes and monitored for up to 60 or 90 days. Finally, a dose corresponding to a volume of one-fifth of the bovine dose volume was chosen for routine vaccination in guinea pigs. Finally, the sampling time and the end of the study were established at 30 post-vaccination days (pvd) for all the viruses. For routine vaccine testing, in all the INTA guinea pig model guidelines submitted to CAMEVET-OIE, in each immunization assay, together with the assessment of unknown vaccine(s) (n = 6 guinea pigs per vaccine), we recommend including two other groups of guinea pigs, one vaccinated with a reference vaccine of known potency (n = 6) provided by the animal health authority and the negative control group (non-vaccinated or vaccinated with placebo, n = 3) (Figure 1.5a). Blood extraction is conducted by saphenous vein puncture or intra-cardiac puncture under anesthesia (final point bleed), following European Centre for the Validation of Alternative Methods (ECVAM) recommendations for animal welfare (Birck et al. 2014). The protocol was approved by the Centro de Investigaciones en ciencias veterinarias y agronómicas,

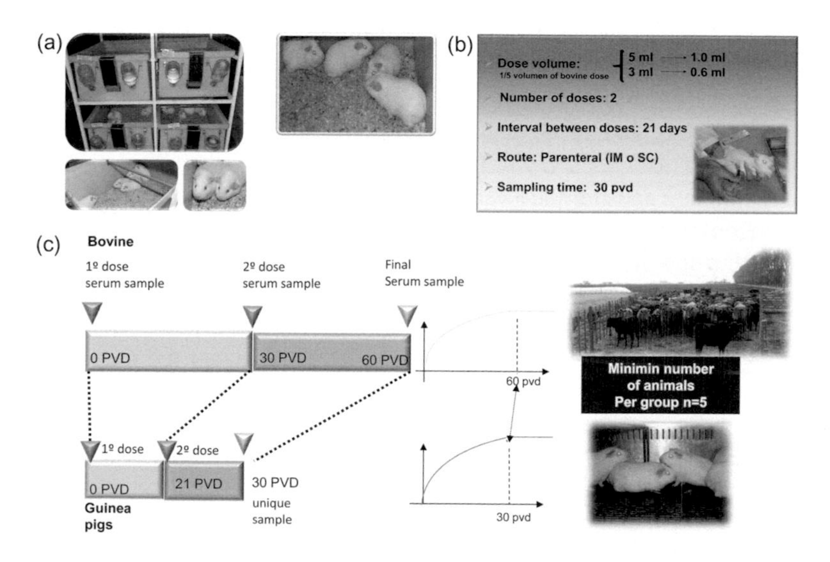

Figure 1.5 (a) Facility to house guinea pigs. (b) INTA guinea pig model for vaccine potency testing: immunization and sampling protocol. (c) For the validation of the lab animal model against the target species. Preliminary studies allow us to establish the dose in guinea pigs to be one-fifth the volume of the bovine dose. Antibody titers in serum samples taken in bovines at 60 pvd were compared with the antibody titers in guinea pig at 30 pvd. The vaccine potency testing in guinea pigs is 1 month faster than in bovines.

Instituto nacional de tecnologías agropecuaria (CICV y A. INTA) Ethical Committee (CICUAE).

Serum Ab responses are tested using specific serological tests like ELISA, viral neutralization (VN), or HIA, depending on the virus included in the vaccine. It is worth mentioning that the guinea pig is a species free of all the bovine viruses included in the vaccines, so they are naturally seronegative for Ab against these viral agents. In addition, based on the results obtained since 2003, where all guinea pig serums obtained at 0 pvd (day of first immunization) were negative, except for PI-3. Some animals showed hemagglutination inhibition titer against bovine PI-3. Thus, for this particular virus, we recommend annual testing of the reproductive animals from the breeding colony to be free of Ab against all viruses. This control allows us to eliminate the initial sampling (0 pvd) of the animals, sampling the vaccinated and control groups only at the end of the test (30 pvd) in order to follow the 3R principle (Akkermans et al. 2020).

Another important point is that the proposed test does not require sophisticated infrastructure or technology but only an animal facility with guinea pigs and a BSL2 laboratory to run the serological tests, such as ELISA, VN, and HIA, all of which are routinely used in virology laboratories.

1.9.2 Why guinea pigs?

Guinea pigs, unlike other laboratory animals such as rats or mice, have the advantage of being bigger in size, thus allowing paired serum sampling, if needed, without risking their lives (Birck et al. 2014). In the guinea pig model, though it is an *in vivo* assay, the number of animals employed (n = 6 per vaccine and 3 negative controls/placebos) and the number of blood extractions are reduced to the minimum (only one sample at 30 days after vaccination is taken).

Furthermore, with the obtained sample volume, the quality for all viral antigens contained in polyvalent vaccines can be assessed. In some case, such vaccines can contain a combination of four or more strains of the following viral agents: bovine herpesvirus 1 and 5; BVDV 1 and 2; RSV; PI-3; and bovine rotavirus G6P[5] and G10P[11]. Some vaccines to prevent bovine neonatal diarrheas also include bovine coronavirus in their formulation. Finally, serological evaluation in the guinea pig model is independent of the type of adjuvant (oil or aqueous) and the amount and quality of inactivated or live attenuated viruses contained in the formulation. Subunit vaccines can also be evaluated. And as we mentioned earlier, in the INTA guinea pig model, only one serum sample is taken, and the animals do not need to be euthanized in the end of the test (Figure 1.5b).

1.10 Experimental design and statistical validation of the INTA guinea pig model for vaccine potency testing

The validation of the model for vaccine potency testing in guinea pigs was conducted following the recommendation of the EMEA Committee for Veterinary Medical Products and ECVAM (Halder et al. 2002; Hendriksen 1999; Taffs 2001; Woodland 2012; Wright 1999). The validation involves three or four steps depending on the virus. Initially, at least two sets of reference vaccines are formulated with known concentrations of the target virus per dose, usually in serial 10-fold concentrations from 10^4 up to a maximum of 10^8, depending on the virus, emulsified in different types of oil adjuvants or adsorbed in aqueous adjuvants (aluminum hydroxide) typically used in this type of vaccine.

The reference vaccines are tested in guinea pigs and bovines in parallel. Bovines are vaccinated with two vaccine doses, 30 days apart. The animals are sampled at 0, 30, and 60 pvd. Usually, the peak or plateau of the Ab response is observed at 60 pvd. With the aim of reducing control times for vaccine release, guinea pigs are immunized also with two doses of one-fifth of the volume of the dose given to bovines at 0 and 21 days and sampled at 30 days. The Ab titer obtained in the lab animal model at 30 days is then correlated with the Ab titer obtained in bovines at 60 days (Figure 1.5c).

The results of these dose–response studies are analyzed using different regression analyses or non-parametric methods, and quality split points are established for the lab animal model and the target species in order to classify the vaccines in terms of the Ab titers induced in guinea pigs at 30 pvd that correspond to the Ab titers that the vaccine induces in cattle at 60 pvd. This is defined as the immunogenicity or potency of the vaccine. According to the viral antigen to be tested, the vaccines are classified as having low, satisfactory, or very satisfactory potency in guinea pigs and will have a corresponding quality or potency in bovines (Figure 1.6a and 1.6b).

To confirm the predictive value of the guinea pig model, a large set of vaccines, if possible including all the formulations present in the market, is tested in both the model and the target species. The concordance analysis between the model and the target species is conducted by using weighted kappa statistics. The weighted kappa gives different weights to disagreements according to the magnitude of the discrepancy, avoiding the weakness of the kappa statistic, which takes no account of the degree of disagreement. Values of weighted kappa from 0.41 to 0.60 indicate moderate agreement; values from 0.61 to 0.80 substantial agreement; and values from 0.81 to 0.99 almost perfect agreement (Vanbelle 2016; Viera

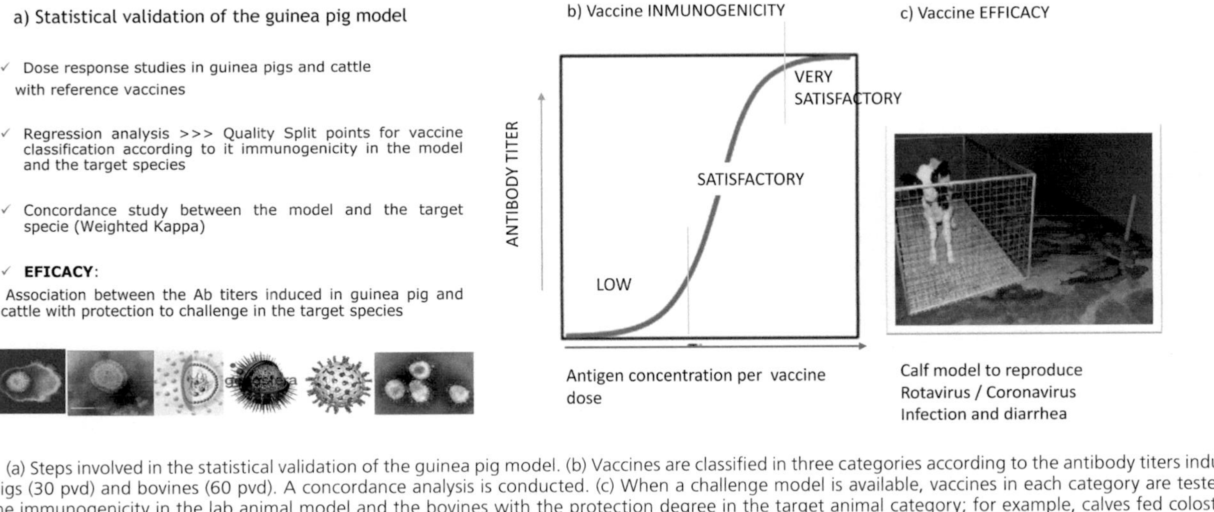

Figure 1.6 (a) Steps involved in the statistical validation of the guinea pig model. (b) Vaccines are classified in three categories according to the antibody titers induced in guinea pigs (30 pvd) and bovines (60 pvd). A concordance analysis is conducted. (c) When a challenge model is available, vaccines in each category are tested to associate the immunogenicity in the lab animal model and the bovines with the protection degree in the target animal category; for example, calves fed colostrum from unvaccinated cows or cows vaccinated with vaccines to prevent calf scour and then challenged with rotavirus and coronavirus.

and Garrett 2005). At least substantial agreement needs to be reached in order to consider the lab animal model an optimal predictive tool for vaccine potency in bovines.

In the case of viral challenge models under controlled conditions in isolation boxes, where the bovines are vaccinated and then challenged with the target virus to establish the efficacy of the vaccine to protect against infection and disease, the Ab titers in guinea pigs and cattle are associated with degrees of protection against infection and/or disease in the target species. In this way, the lab animal model can also predict not only the immunogenicity of the vaccine in the target species but also the degree of protection against infection and disease (Figure 1.6c). In the present work, the guinea pig model was validated against protection to challenge for IBR, BVDV, and rotavirus.

2
Guinea Pig Model to Test the Potency of IBR Vaccines

Development and statistical validation of a guinea pig model and its associated serology assays as an alternative method for potency testing of bovine herpesvirus vaccines.

2.1 Introduction

Bovine herpesvirus 1 (BoHV-1) is the etiological agent of infectious bovine rhinotracheitis/infectious pustular vulvovaginitis (IBR/IPV) (OIE 2021), a disease of domestic and wild cattle that causes a wide range of clinical signs, including rhinotracheitis, vulvovaginitis, infectious pustular balanoposthitis, conjunctivitis, abortion, enteritis, and encephalitis (Fulton 2020; Loy, Clothier, and Maier 2021).

The respiratory form of the disease includes coughing, nasal discharge, and conjunctivitis. Signs can range from mild to severe, depending on the presence of secondary bacterial pneumonia with the development of dyspnea. In the absence of bacterial pneumonia, recovery generally occurs 4–5 days after the onset of signs. After respiratory and genital infection, BoHV-1 becomes latent in the neural ganglia. Stress can induce reactivation of the latent infection, and the virus may be shed intermittently. Infection elicits an antibody response and a cell-mediated immune response within 7–10 days. Neutralizing antibodies may persist 5 years after infection, but re-stimulations (reactivation or vaccination) are needed to keep titers at detectable levels by the viral neutralization technique (Fulton 2020; Loy, Clothier, and Maier 2021). On the contrary, total antibodies evaluated by enzyme-linked immunosorbent assay (ELISA), remain detectable for life (Parreño et al. 2010b).

Because virus latency is a normal sequel to BoHV-1 infection, and antibody (Ab) response after infection seems to be lifelong, any seropositive animal should be considered as a potential carrier and intermittent shedder of the virus, with the exception of young calves with passive maternal Ab and non-infected cattle vaccinated with killed vaccines (Brock et al. 2021; Romera et al. 2014).

DOI: 10.1201/9781003373148-2

In general, vaccines prevent the development of severe clinical symptoms and reduce the shedding of virus after infection, but they do not prevent infection. Several eradication campaigns with and without vaccination (mandatory and/or voluntary) are being carried out in Europe (Brock et al. 2021; Iscaro et al. 2021; Maresca et al. 2018). Norway, Finland, Sweden, Austria, Denmark, Switzerland, and various regions of Italy and Germany have eradicated the infection. In the rest of the world, infection is endemic and with high prevalence.

Various attenuated and inactivated BoHV-1 vaccines are currently available in the market. In Argentina and Uruguay, the only authorized vaccines are inactivated ones. Vaccines contain strains of the virus, generally replicated during multiple passages in cell culture. Inactivated vaccines contain high levels of inactivated virus or portions of the virus particle (glycoproteins) supplemented with an adjuvant to stimulate an adequate immune response. Inactivated vaccines are administered intramuscularly or subcutaneously. Marker or DIVA (Differentiating Infected from Vaccinated Animals) vaccines have long been available in various countries (Aida et al. 2021). These marker vaccines are based on deletion mutants in a subunit of the virion, for example, glycoprotein E (gE). This type of vaccine is used in Europe in countries that carry out eradication programs with vaccination campaigns (Iscaro et al. 2021; Maresca et al. 2018). In endemic countries, intensive vaccination programs may reduce the prevalence of infected animals and improve productive indexes (Loy, Clothier, and Maier 2021).

For the approval of vaccines containing IBR, international control organizations (Animal and Plant Health Inspection Service [APHIS], USA; the European Medicines Agency Committee for Veterinary Medicinal Products [EMEA-CVMP], EU; World Organisation for Animal Health [OIE]) require potency and efficacy assays in the target species, which imply vaccination and challenge of susceptible and seronegative bovines (9CFR 113.216 2014; OIE 2021). Once the product is approved, the quality of each batch to be released must be controlled by a potency test that determines product immunogenicity in bovines or other laboratory animal model (***in vivo*** test). Some agencies, for example the United States Department of Agriculture (USDA), allow ***in vitro*** potency tests using a parallel line assay and a validated reference vaccine (9CFR 113.216 2014). The ***in vitro*** potency test must be statistically validated and show an acceptable agreement when compared with the potency test in the target species. It is also strongly desirable that the model be validated as a predictive tool of the degree of protection that the vaccine will provide against the viral shed in seronegative bovines (Parreño et al. 2010a).

A BoHV-1 vaccine must prevent the development of severe clinical signs and markedly reduce virus shedding after experimental challenge. Bovine trials are cumbersome, expensive, and time consuming, particularly in countries like Argentina, where BoHV-1 as well as other viral infections is endemic (Moore et al. 2003). The difficulty in finding seronegative bovines from BoHV-1-free herds to be used in vaccine potency tests poses the need for developing standardized and harmonized tests in laboratory animals. The availability of a laboratory animal model would enable the regulatory authority and vaccine manufacturers to carry out batch-to-batch release tests on a routine basis in a less time-consuming and less expensive way.

Although some vaccine manufacturers have reported the use of guinea pigs as an internal quality test to evaluate their vaccines, a validated method for vaccine potency testing in laboratory animals possessing a demonstrated concordance with the target species is not yet available. Such a properly validated vaccine potency test, especially designed for combined vaccines, including inactivated viruses, is also required in the US and the EU and could be globally used to control viral vaccines applied in cattle (Akkermans et al. 2020; Hendriksen 1999; Taffs 2001).

Although several ELISA tests were developed to determine BoHV-1 Ab and probed to be more sensitive and specific than the viral neutralization (VN) test, the latter is still considered the gold standard technique used for vaccine potency testing (OIE 2021).

In the present chapter, we report the statistical validation of a guinea pig model as a method for potency testing of inactivated IBR vaccines. As explained in Chapter 1, the validation involved the study of the kinetics of the Ab response in the animal model and the target species, a regression analysis applied to the dose–response curve to define categories for vaccine qualification, and a concordance analysis between the laboratory animal model and the natural host, confirmed with a BoHV-1 experimental challenge in the latter to correlate the Ab titer with protection in the target species.

2.2 Dose–response analysis: kinetics of the antibody response and selection of the sampling time points for guinea pigs and bovines

Three sets of vaccines consisting of combined water-in-oil vaccines containing increasing concentrations of BoHV-1 covering a range from 10^5 to 10^7 50% tissue culture infective dose ($TCID_{50}$)/dose and two vaccines with 10^8 $TCID_{50}$/dose, together with fixed concentrations of parainfluenza virus type 3 (PI-3) and bovine viral diarrhea virus (BVDV), were formulated. Viral antigens were inactivated using binary ethylenimine (BEI) (Bahnemann 1990), and vaccines were formulated using pilot-scale equipment. The adjuvant used in the formulation consisted of a mix of 0.67% Polysorbate 80, 2.1% Sorbitan 80, Monooleate, and 57.9% mineral oil, in a 60:40 oil:water proportion. Each vaccine set was formulated with a different dilution of the same bulk of antigen (Ag) in order to obtain the desired Ag concentration. The vaccines were formulated thereafter and were applied following the same time intervals and dose volumes as commercial vaccines. These vaccines of known antigen concentration are referred to as "reference vaccines" and were used to estimate the dose–response calibration curve in bovine calves and guinea pigs. Each set of reference vaccines was tested in two independent experiments in guinea pigs and two independent trials in bovine calves (Table 2.1).

In the initial stage of development, samples were taken at different time points, and the kinetics of the antibody response (mean ELISA and VN antibody titers) induced by this first set of reference vaccines (calibration vaccines) were evaluated until 60 days post-vaccination (dpv) in guinea pigs and 90 dpv in calves. In all cases, vaccines were administered in parallel, in groups of a minimum of five guinea pigs and five bovines. The minimum number of repetitions (animals per group; $n = 5$) was calculated to achieve a statistical power of at least 83% to discriminate between vaccines containing BoHV-1 concentrations differing in 1 $\log_{10}$.

The dose–response experiments in bovine calves were conducted in six different herds, using a total of five calves per treatment group. A first set of vaccines was tested in two independent trials involving seronegative calves from two BoHV-1-free farms without history of previous vaccination or epidemiological evidence of BoHV-1 infection (Figure 2.1). A second set of vaccines was also evaluated in two bovine trials. The first experiment was carried out using selected seronegative calves from an endemic herd and the second in a non-vaccinated herd but with a history of natural infection with BoHV-1 and including seronegative and seropositive calves (pre-vaccination Ab titer VN: 0.3–1.5; ELISA: 0.3–2.8). Finally, a third set of vaccines was tested in another two BoHV-1-free farms.

Blood samples for serum extraction were collected by puncture of the jugular or coccygeal vein. Bovines were sampled at 0, 30, and 60 post-vaccination days (pvd). Control groups included placebo and non-vaccinated animals. Groups of calves that received two doses of placebo formulated with culture media (without virus) emulsified in oil adjuvant were assayed

Table 2.1 IBR Vaccines Tested in Parallel in Guinea Pigs and Bovines for the Model Validation

Type of vaccine	Syndrome	Vaccine composition (viral antigens	BoHV-1 concentration ($TCID_{50}$/dose)	Number of vaccines tested	Number of vaccinated bovines	Number of vaccinated guinea pigs	Number of comparative assays[a]
Calibration vaccines for the dose–response curve[b] (11 vaccines; 20 comparative assays)[c]	Respiratory	IBR-BVDV-PI-3	1×10^8	2	20	20	4
			1×10^7	3	30	30	6
			1×10^6	3	30	36	6
			1×10^5	3	30	30	4
Reference vaccines for concordance analysis[d] (n = 7 vaccines; 12 assays)[c]	Reproductive	IBR-BVDV	10^7	3	70	51	8
		DIVA gE–	10^7	1	5	5	1
	Respiratory	IBR-BVDV-PI-3	10^8	1	5	5	1
			$10^{7.5}$	1	5	5	1
			10^7	1	5	5	1
Commercial vaccines[a] (n = 18 vaccines; 22 assays)[c]	Respiratory	IBR-BVDV-PI-3-BRSV	Unknown	3	30	30	5
		IBR-BVDV-PI-3		5	29	40	6
	Reproductive	IBR-BVDV		4	35	19	4
	Conjunctivitis	IBR		2	15	11	2
	Multi-purpose	IBR-BVDV-PI-3-RV		4	35	35	5

(*Continued*)

Table 2.1 (Continued) IBR Vaccines Tested in Parallel in Guinea Pigs and Bovines for the Model Validation

Type of vaccine	Syndrome	Vaccine composition (viral antigens	BoHV-1 concentration ($TCID_{50}$/dose)	Number of vaccines tested	Number of vaccinated bovines	Number of vaccinated guinea pigs	Number of comparative assays[a]
Placebo				8	58	44	8
Total vaccinated				44	402	366	62
Non-vaccinated					151	131	23
Total				44	553	497	85

[a] Oil commercial vaccines included water-in-oil and double water-oil-water emulsions.

[b] Three sets of water-in-oil vaccines with decreasing concentrations of BEI-inactivated BoHV-1 emulsified in oil adjuvant were prepared except for the version containing 10^8 $TCID_{50}$/dose of which only two vaccine were tested. Each vaccine set was evaluated in two independent experiments in guinea pigs and two independent field trials in bovines. Each group included five guinea pigs and five bovines. except for vaccines with 10^6, where 6 animals were added to increase the number of replicates in the concentration considered, under these conditions, the detection limit of the model. Groups of 3–5 animals, vaccinated with placebo and non-vaccinated, were included in each assay as negative controls.

[c] Some vaccines were tested in more than one occasion to evaluate stability through time, generating a higher number of comparison lines.

[d] Reference vaccines: second set of vaccines of known antigen concentration (potency) prepared under industrial conditions used for concordance analysis included water-in-oil emulsions.

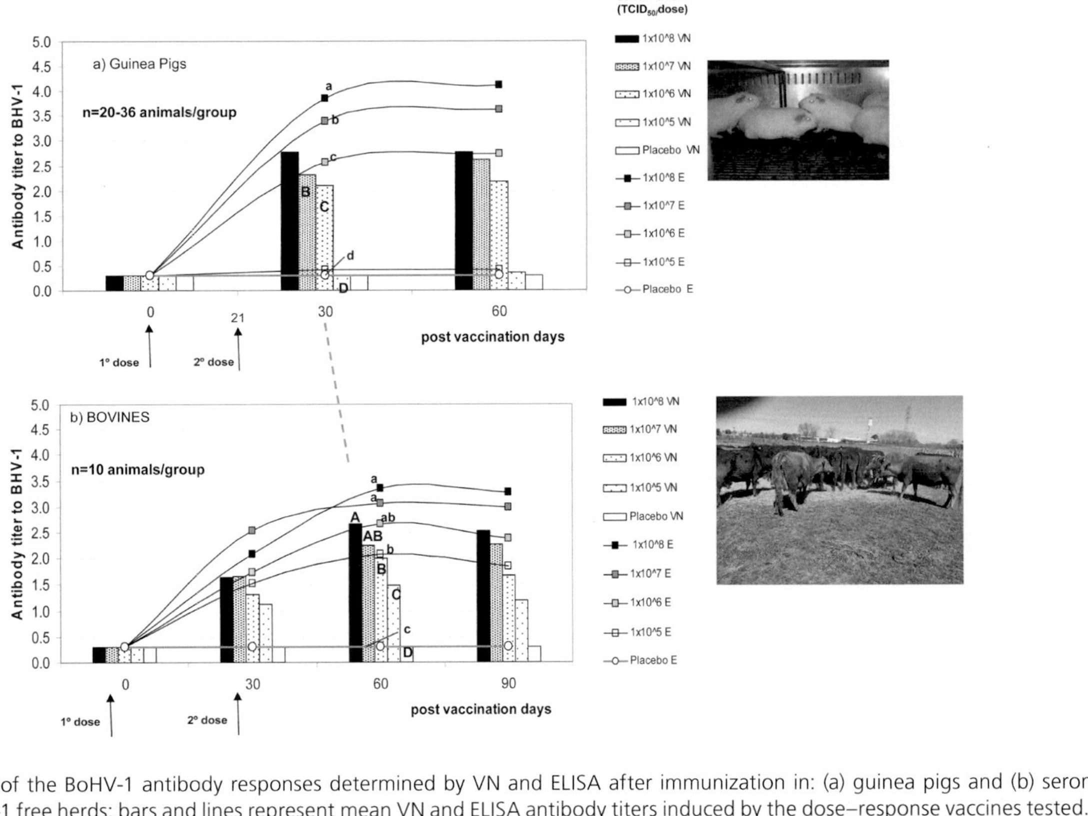

Figure 2.1 Kinetic of the BoHV-1 antibody responses determined by VN and ELISA after immunization in: (a) guinea pigs and (b) seronegative bovines from BoHV-1 free herds; bars and lines represent mean VN and ELISA antibody titers induced by the dose–response vaccines tested. respectively. Arrows indicate vaccination time. Bars/lines. at 30 dpv. with different upper/lower case letters indicate statistical differences among mean VN/ELISA antibody titers induced by the vaccine tested (mixed model for repeated measures and Bonferroni method for multiple comparison. $p < 0.05$) (Di Rienzo, Guzman, and Casanoves 2002).

on eight occasions, including the dose–response trials, in which the animals were sampled until 90 pvd. A bovine trial was only considered valid and included in the statistical analysis if no seroconversion was detected in the non-vaccinated and placebo control groups. Most experiments were carried out blinded (veterinarian and laboratory technician). Seroconversion was defined as a fourfold increase in antibody titer for both ELISA and VN.

Antibody responses were determined by VN and ELISA. The VN assay used in this study, considered as the "gold standard" or "traditional test", was performed as described in Manual of Terrestrial Animals (OIE 2021) and adjusted to meet the recommendations given by the OIE and American Code of Federal Regulations (CFR). An ELISA for the quantification of total antibody to BoHV-1 in bovine and guinea pigs was specifically validated for the model (Table 2.2) (Parreño et al. 2010b).

The three sets of reference vaccines formulated with increasing concentrations of BoHV-1 per dose (Table 2.1) were tested in two independent assays in guinea pigs and two independent trials in bovine calves, giving a total of six points for each BoHV-1 concentration. Only two sets of vaccines containing the highest concentration (1×10^8 $TICD_{50}$/dose) were tested, giving a total of four replicates.

The curves of the kinetics of the antibody response to BoHV-1 after vaccination in both species were analyzed using a mixed model for repeated measures and Bonferroni method for multiple comparisons (Bretz et al. 2011). This analysis enabled the determination of the earliest time point for animal sampling, corresponding to the peak or plateau Ab titer that also showed acceptable discriminatory capacity.

Table 2.2 ELISA Validation Parameters for the Detection of IgG Antibodies in Bovines and IgG in Guinea Pigs for IBR Vaccine Potency Testing

ELISA validation parameters	IgG anti-IBR (guinea pigs)	IgG1 anti-IBR (bovine)
Cut-off	40%PP[a]	40%PP[a]
Sensitivity	95%	89.74%
Specificity	100%	100%
Intermediate precision (3 years)	CV[b] = 22.2%	CV[b] = 23.1%
Accuracy (ROC[c] analysis)	99.8%	99%

[a] PP = percentage of positivity compared with a positive control run in every plate and every run

[b] CV = coefficient of variation <25% recommended by OIE (Wright 1999).

[c] ROC = receiver operating characteristic.

At 30 pvd, after receiving two doses of vaccine, the guinea pigs developed a strong Ab response to BoHV-1. A clear dose-dependent behavior according to the Ag concentration of the vaccine was observed. The antibody titers remained high and constant until 60 pvd. For the lab animal model, 30 pvd corresponded to the peak Ab titer and the earliest time point of maximum discrimination among vaccines differing by 1 $\log_{10}$ in their antigen concentration (Figure 2.1a).

After vaccination, all groups of bovines developed Ab response to BoHV-1, and 60 pvd was the time point corresponding to the peak of BoHV-1 Ab titers. At 90 pvd, the response remained in a plateau (ELISA) or started to decrease (VN) (Figure 2.1b). The shape of the kinetic curves of the Ab response for both species was similar. The Ab titers detected in guinea pigs at 30 pvd were comparable to those detected in bovines at 60 pvd. Thus, those time points were selected as the optimal moments for sampling and were used as the unique bleeding point for further concordance studies and for future routine use of the guinea pig model as a method for vaccine potency testing.

2.3 Dose–response analysis: detection limit for comparison between the model and the target species

In a second stage of validation, at the time points selected in the previous stage, the ability of the laboratory animal model and bovine calves to significantly discriminate among vaccines formulated with antigen concentrations of 1 $\log_{10}$ difference was evaluated. At this stage, the minimal antigen concentration inducing a detectable Ab response by ELISA and VN was determined (detection limit or minimal immunogenicity dose). Each individual assay/trial using a group of five animals was analyzed by one-way analysis of variance (ANOVA) followed by DGC multiple comparison test (Di Rienzo, Guzman, and Casanoves 2002). The Kruskal–Wallis non-parametric rank sum test was used when the assumptions of normality and/or homoscedasticity were not met. In all cases, the significance level was established at 5%. The statistical analysis was conducted using SAS statistical software (SAS Institute Inc.).

At 30 pvd, the guinea pig model was able to discriminate between vaccines formulated with 10^7 $TCID_{50}$/dose of BoHV-1 or higher from vaccines containing 10^6 TCID50/dose and 10^5 $TCID_{50}$/dose by ELISA. In all the experiments, vaccines with 10^7 $TCID_{50}$/dose of BoHV-1 induced mean Ab titers higher than 3.00, while vaccines containing 10^6 TCID50/dose of BoHV-1 induced mean Ab titers lower than 3.00. Vaccines formulated with

10^5 $TCID_{50}$/dose did not induce an Ab response, indicating that under the conditions used, this concentration was below the detection limit of the model and the serology test used. The global analysis considering the total number of guinea pigs vaccinated with each Ag dose showed that the guinea pig model was able to significantly differentiate among vaccines formulated with BoHV-1 concentrations differing by 1 log10 (Figure 2.1a).

The groups of calves vaccinated with decreasing doses of BoHV-1 also showed a dose–response effect. In all cases, vaccines formulated with 10^6 $TCID_{50}$/dose or higher induced an Ab response, while vaccines formulated with 10^5 $TCID_{50}$/dose did not induce a detectable Ab response during the entire period of the trial (Figure 2.1b).

The Ab response measured by VN, the traditional technique, consistently yielded lower Ab titers than those obtained by ELISA for both species. No neutralizing Ab responses were detected in the groups of animals vaccinated with 10^5 $TCID_{50}$ of BoHV-1/dose, indicating that the limit of detection was similar for both techniques (Parreño et al. 2010a) (figure 3; supplementary data Tables 1 and 2).

2.4 Repeatability and reproducibility of the target species and the lab animal model

Statistical analysis for the vaccines tested for each antigen concentration, considering all the experiments and trials involved in the dose–response study, was performed as a nested analysis of variance (nested ANOVA). The applied model allowed quantification of the relative contribution of the different sources of variation associated with the repeatability and reproducibility of the host species and the guinea pig model to evaluate vaccine potency. The repeatability was expressed as the coefficient of variation (CV) or the overall relative variation between individual bovine calves and guinea pigs within a group, and the reproducibility was expressed as the CV or overall relative variation between different groups vaccinated with vaccine containing the same BoHV-1 concentration (Pryseley et al. 1999).

The guinea pig model gave highly consistent results of Ab titers by both ELISA and VN among different assays for the same Ag concentrations in the range from 10^6 to 10^8 $TCID_{50}$/dose, giving acceptable repeatability and reproducibility. The CV associated with the reproducibility of the test was lower than 10% for all the antigen concentrations tested and tended to zero (the variation among different assays was irrelevant) for vaccines formulated with 10^7 $TCID_{50}$/dose. The target species also gave consistent results for testing the potency of vaccines containing 10^7 and 10^8 $TCID_{50}$/

Table 2.3 Repeatability and Reproducibility of the Guinea Pig Model versus the Target Species Expressed as CV

BoHV-1 concentration ($TCID_{50}$/ml)	Bovine		Guinea pig	
ELISA	Repeatability	Reproducibility	Repeatability	Reproducibility
1×10^8	14.3%	14.6%	8.9%	1.6%
1×10^7	13.8%	18.4%	20.6%	→0%[a]
1×10^6	15.3%	21.52%	18.4%	9.9%
1×10^5	na	na	na	na
VN	Repeatability	Reproducibility	Repeatability	Reproducibility
1×10^8	18.0%	11.7%	8.8%	7%
1×10^7	18.8%	16.9%	17.9%	9.8%
1×10^6	25.4%	29.3%	14.3%	11.9%
1×10^5	na	na	na	na

[a] CV tended to zero. indicating that the variation among assays was irrelevant.

dose (CV associated with the reproducibility <20%) for both techniques (Table 2.3).

2.5 Regression analysis to establish the cut-off for vaccine classification

The VN and ELISA mean Ab titers of the groups of bovine calves and guinea pigs immunized with the calibration vaccines formulated for the dose–response assay, at the previously selected time points (30 days in guinea pigs and 60 days in bovines), were analyzed by regression analysis. The lower limits of the 90% prediction intervals were used as classification criteria to establish a range of vaccine quality acceptance in terms of immunogenicity (potency) in both species. The calculated split points were further used for vaccine classification into three categories: "unsatisfactory", "satisfactory", and "very satisfactory". The protection rate of vaccines representing these categories was evaluated by BoHV-1 experimental challenge in the natural host.

All the dose–response experiments conducted in guinea pigs and 5 out of 6 trials in calves were considered valid, giving a total of 22 groups for the laboratory animal model and 20 groups for the host species to be included in the regression analysis, as depicted in Figure 2.2. The mean antibody titers determined by ELISA and VN were directly related to the BoHV-1 concentration in the vaccines. For both species and techniques, the mathematical model that best fitted the data was a polynomic linear model of a second

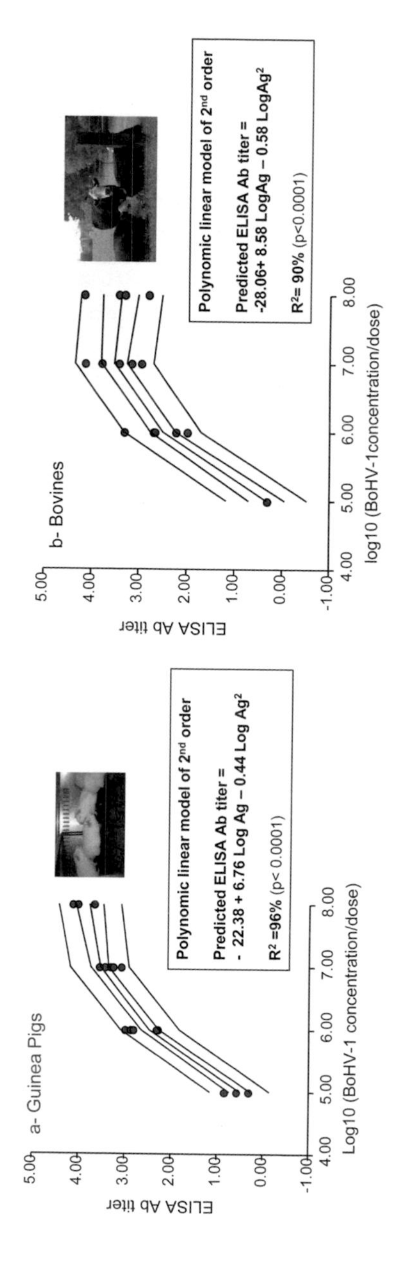

Figure 2.2 Regression analysis and estimation of "split points" for vaccine classification in (a) guinea pigs and (b) bovines by ELISA.

Table 2.4 Vaccine Classification Criteria by ELISA

	Vaccine potency (ELISA)		
Species	**Non-satisfactory**	**Satisfactory**	**Very satisfactory**
Guinea pig	$\bar{y} < 1.93$	$1.93 \leq \bar{y} < 3.02$	$3.02 \leq \bar{y}$
Bovine	$\bar{Y} < 1.69$	$1.69 \leq \bar{Y} < 2.72$	$2.72 \leq \bar{Y}$

Cut-offs determined by ELISA expressed as the $\log_{10}$ of the reciprocal of the analyzed serum dilution that has a positive result in the assay. Mean Ab titer of groups of five guinea pigs, evaluated 30 days post-vaccination (dpv), and groups of five seronegative bovines evaluated at 60 dpv. Bovines receive two doses of vaccine with a 30-day interval. following vaccine manufacturer's recommendations, and are sampled at 0 and 60 dpv. This latter point corresponded to the peak or plateau of Ab titers reached by aqueous or oil vaccines, respectively. Guinea pigs receive two doses of vaccine (1/5 the volume of the bovine dose) with a 21-day interval and are sampled at 0 and 30 dpv. The two-dose regimen chosen in the lab animal model allows detection of the immune response induced by vaccines of low potency. The 21-day interval between doses was adopted in order to obtain a curve of Ab kinetic response similar to that obtained in bovines, but in a shorter period of time, providing a faster alternative method for vaccine potency testing than the one conducted in bovines.

order of magnitude. The mathematical model was able to predict ELISA Ab titers with $R^2 = 96\%$ for guinea pigs and $R^2 = 90\%$ for bovines (Figure 2.2a and 2.2b). In the case of VN, the model was able to predict VN Ab titers for guinea pigs with $R^2 = 90\%$ and $R^2 = 93\%$ for bovines (Figure 2.4).

The 90% prediction intervals were calculated using the lower limit of Ab titer induced for vaccines formulated with 10^6 and 10^7 $TCID_{50}$/dose of BoHV-1 as quality split points. These limits represent the minimum mean Ab titer that must be induced in a group of a minimum of five guinea pigs and five calves after vaccination with water-in-oil vaccines containing those Ag concentrations with 95% coverage. As depicted in Figures 2.2 and 2.3, and detailed in Tables 2.3 and 2.4, the estimated split points allowed the classification of the vaccines in the guinea pig model and the target species by the two techniques as having "unsatisfactory", "satisfactory", or "very satisfactory" potency.

Ab titers determined by ELISA as higher than 3.02 in guinea pigs and 2.72 in bovines were associated with very satisfactory potency of vaccines. Vaccines inducing Ab titers between 3.02 and 1.93 in guinea pigs and between 2.72 and 1.69 in bovines were satisfactory. Vaccines that induced Ab titers lower than 1.93 in guinea pigs and 1.69 in bovines were considered non-satisfactory for commercialization (Table 2.4).

Neutralizing Ab titers higher than 2.05 in guinea pigs and 1.96 in bovines were associated with very satisfactory potency of vaccines. Vaccines

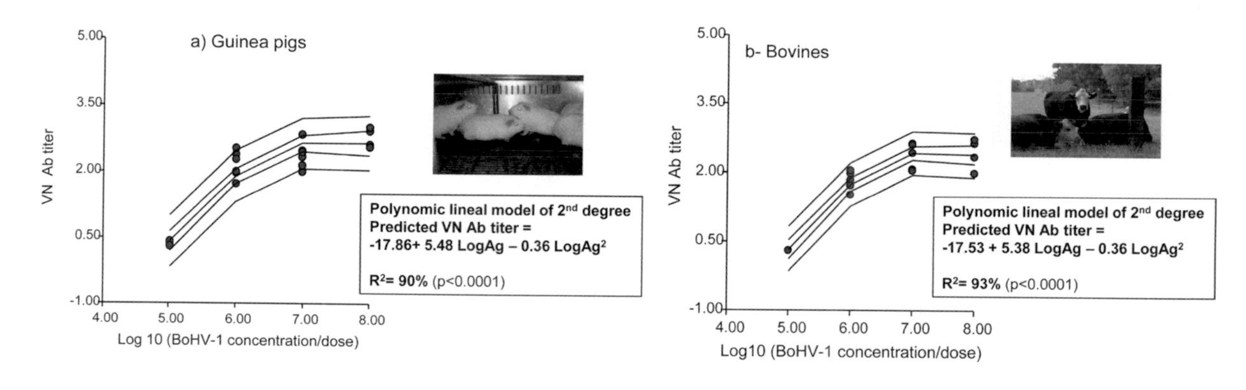

Figure 2.3 Regression analysis and estimation of "split points" for vaccine classification in (a) guinea pigs and (b) bovines by VN.

Table 2.5 Vaccine Classification Criteria by VN

	Vaccines potency viral neutralization (VN)		
Species	**Non-satisfactory**	**Satisfactory**	**Very satisfactory**
Guinea pig	$\bar{y} < 1.31$	$1.31 \leq \bar{y} < 2.05$	$2.05 \leq \bar{y}$
Bovine	$\bar{Y} < 1.27$	$1.27 \leq \bar{Y} < 1.96$	$1.96 \leq \bar{Y}$

Cut-offs determined by VN expressed as Ab neutralizing titers calculated by the Reed and Muench method. Mean Ab titer of groups of five guinea pigs, evaluated at 30 dpv, and groups of five seronegative bovines, evaluated at 60 dpv. Bovines receive two doses of vaccine at a 30-day interval and are sampled at 0 and 60 dpv. Guinea pigs receive two doses of vaccine (1/5 the volume of the bovine dose) at a 21-day interval and are sampled at 30 dpv.

inducing Ab titers between 2.05 and 1.31 in guinea pigs and between 1.96 and 1.27 in bovines were satisfactory. Vaccines that induced Ab titers lower than 1.31 in guinea pigs and 1.27 in bovines were considered non-satisfactory and therefore unsuitable for commercialization (Table 2.5).

2.6 Concordance analysis between bovines and guinea pigs for vaccine classification

To evaluate the performance of the selected split point for vaccine classification and the agreement between the guinea pig model and the target species, a concordance analysis was conducted that involved the evaluation of 22 vaccines in a total of 497 guinea pigs and 553 male and female beef calves (Aberdeen Angus, Hereford, and their crossbreeds) 6–12 months old.

Vaccination trials were conducted in 12 beef farms located in Buenos Aires, Argentina. Herds without previous history of vaccination against BoHV-1 were selected. As BoHV-1 infection is endemic in Argentina, vaccines were evaluated in BoHV-1-seronegative animals from BoHV-1-free herds, and also in seronegative and seropositive calves from BoHV-1-endemic herds, in order to consider the variability of the real target population. In the trials conducted in BoHV-1-endemic farms, the number of positive and negative animals was randomly distributed in each treatment group so as to initiate the study with statistically similar pre-vaccination mean Ab titers and variances among groups. In every trial, bovines were vaccinated with 2 doses of vaccine, 30 days apart, as recommended on the label of each of the 22 commercial vaccines tested. Vaccines were administered by the subcutaneous route with doses of 5 or 3 ml according to the manufacturer's recommendations. In addition, in order to have vaccinated and non-vaccinated groups exposed to similar natural conditions, a group of non-vaccinated calves was included in every bovine trial conducted in each of the 12 farms that participated in this study.

In a first analysis, a total of 63 pairs of groups of vaccinated calves and guinea pigs were tested. This first analysis used data generated from 20 groups of animals vaccinated with the calibration vaccines used to estimate the dose–response curve and another 12 assays including a second set of 7 combined water-in-oil immunogens produced following the industrial outline of production ($\geq 10^7$ $TCID_{50}$/dose). These vaccines of known antigen concentration and consequently, of known potency were also assigned as gold standard vaccines and included three inactivated IBR-BVDV oil vaccines used for prevention of bovine reproductive syndrome, a gE-DIVA vaccine developed in Argentina (Romera et al. 2014), and three multivalent respiratory vaccines, all emulsified oil adjuvant (0.67% Polysorbate 80, 2.1% Sorbitan 80 Monooleate, and 57.9% mineral oil; 60:40 oil:water proportion). Eight placebo groups and 23 non-vaccinated groups were considered as the negative control groups.

To avoid bias, in a second analysis, the calibration vaccines used to set the split points ($n = 20$) and most of the non-vaccinated groups were subtracted ($n = 16$), and commercial vaccines were added instead. Thus, 41 pairs of groups were included: the 12 assays with reference vaccines, and 22 commercial vaccines and their corresponding negative control groups ($n = 7$). The commercial vaccines included in this analysis are products marketed for the prevention of viral conjunctivitis and respiratory and reproductive disease of cattle, from various manufacturers, and contain unknown titers of inactivated BoHV-1. In all cases, vaccines were administered in parallel to 5–10 guinea pigs and bovine calves per group (Table 2.1).

Concordance between the vaccine quality predicted by the guinea pig model and that obtained in the natural host was estimated by the weighted kappa coefficient. The weighted kappa gives different weights to disagreements according to the magnitude of the discrepancy, avoiding the weakness of the kappa statistic, which takes no account of the degree of disagreement. Values of weighted kappa from 0.41 to 0.60 indicate moderate agreement; values from 0.61 to 0.80 substantial agreement; and values from 0.81 to 0.99 almost perfect agreement (Viera and Garrett 2005).

Concordance between the vaccine quality predicted by the guinea pig model and that obtained in the natural host by ELISA and VN is shown in Tables 2.6 and 2.7. The initial analysis including 63 groups of animals immunized with vaccines of known antigen concentration and placebo or non-vaccinated control groups assigned as "reference groups" gave almost perfect agreement in vaccine classification between the model and the target species by both techniques (Table 2.6a and b).

When excluding the pairs of groups used in the model estimation (20 vaccines used to estimate the calibration curve, 8 placebos, and 18 out of 23 non-vaccinated control groups) and adding instead 22 commercial vaccines of unknown potency, the weighted kappa values were lower (Table 2.7). Nevertheless, the agreement remained almost perfect for ELISA and

Table 2.6 Concordance between the Guinea Pig Model and Bovines Analyzed by ELISA (a) and VN (b) for the Analysis of Vaccines of Known Potency[a]

(a) ELISA; *n* = 63

Guinea pig \ Bovine	Unsatisfactory ($\overline{X}$ < 1.69)	Satisfactory (1.69 ≤ $\overline{X}$ < 2.72)	Very satisfactory (2.72 ≤ $\overline{X}$)
Unsatisfactory ($\overline{X}$ < 1.93)	**36**	2	0
Satisfactory (1.93 ≤ $\overline{X}$ < 3.02)	0	**5**	3
Very satisfactory (3.02 ≤ $\overline{X}$)	0	1	**16**

Weighted kappa: 0.894; asyntotic standard error (ASE) = 0.041; 95% confidence interval (CI) 0.813–0.974; $p < 0.0001$; almost perfect agreement.

(b) VN; *n* = 63

Guinea pig \ Bovine	Unsatisfactory ($\overline{X}$ < 1 27)	Satisfactory (1.27 ≤ $\overline{X}$ < 1.96)	Very satisfactory (1.96 ≤ $\overline{X}$)
Unsatisfactory ($\overline{X}$ < 1.31)	**37**	0	1
Satisfactory (1.31 ≤ $\overline{X}$ < 2.05)	1	**4**	1
Very satisfactory (2.05 ≤ $\overline{X}$)	0	3	**16**

Weighted kappa: 0.78; ASE = 0.050; 95% CI 0.777–0.971; $p < 0.0001$; almost perfect agreement. Results are highlighted in bold.

[a] The analysis included 20 assays corresponding to the calibration vaccines used in 8 placebos; 23 non-vaccinated groups and 12 assays including vaccines of known potency were also assigned as reference vaccines.

Table 2.7 Concordance between the Guinea Pig Model and Bovines for the Analysis of Vaccines of Known Potency and Commercial Vaccines of Unknown Potency[a]

(a) ELISA; *n* = 41

Guinea pig \ Bovine	Unsatisfactory ($\overline{X}$ < 1.69)	Satisfactory (1.69 ≤ $\overline{X}$ < 2.72)	Very satisfactory (2.72 ≤ $\overline{X}$)
Unsatisfactory ($\overline{X}$ < 1.93)	**12**	0	0
Satisfactory (1.93 ≤ $\overline{X}$ < 3.02)	0	**3**	3
Very satisfactory (3.02 ≤ $\overline{X}$)	0	2	**21**

Weighted kappa: 0.865; ASE = 0.060; 95%CI 0.748–0.982; $p < 0.0001$; almost perfect agreement. Results are highlighted in bold.

(b) VN; *n* = 41

Guinea pig \ Bovine	Unsatisfactory ($\overline{X}$ < 1.27)	Satisfactory (1.27 ≤ $\overline{X}$ < 1.96)	Very satisfactory (1.96 ≤ $\overline{X}$)
Unsatisfactory ($\overline{X}$ < 1.31)	**12**	0	1
Satisfactory (1.31 ≤ $\overline{X}$ < 2.05)	2	**10**	1
Very satisfactory (2.05 ≤ $\overline{X}$)	0	4	**11**

Weighted kappa: 0.761; ASE = 0.082; 95%CI 0.601–0.921; $p < 0.0001$; substantial agreement. Results are highlighted in bold.

[a]The analysis comprised 12 assays including vaccines of known antigen concentration (known potency) referred to as "reference vaccines" and 22 assays including commercial vaccines of unknown potency and their corresponding negative control groups ($n = 7$).

substantial for VN, according to the criteria established by Viera et al. (Viera and Garrett 2005).

Finally, when only the 22 commercial vaccines were analyzed, the concordance between the guinea pig model and the host species was almost perfect for both techniques (Table 2.8).

Table 2.8 Concordance between the Guinea Pig Model and Bovines for the Analysis of Commercial Vaccines[a]

(a) ELISA; $n = 22$

Guinea pig \ Bovine	Unsatisfactory ($\overline{X} < 1.69$)	Satisfactory ($1.69 \leq \overline{X} < 2.72$)	Very satisfactory ($2.72 \leq \overline{X}$)
Unsatisfactory ($\overline{X} < 1.93$)	**5**	0	0
Satisfactory ($1.93 \leq \overline{X} < 3.02$)	0	**1**	1
Very satisfactory ($3.02 \leq \overline{X}$)	0	2	**21**

Weighted kappa: 0.8842; ASE = 0.083; 95%CI 0.722–1.00; $p < 0.0001$; almost perfect agreement. Important results are highlighted in bold.

(b) VN; $n = 22$

Guinea pig \ Bovine	Unsatisfactory ($\overline{X} < 1.27$)	Satisfactory ($1.27 \leq \overline{X} < 1.96$)	Very satisfactory ($1.96 \leq \overline{X}$)
Unsatisfactory ($\overline{X} < 1.31$)	**5**	0	0
Satisfactory ($1.31 \leq \overline{X} < 2.05$)	1	**8**	0
Very satisfactory ($2.05 \leq \overline{X}$)	0	2	**6**

Weighted kappa: 0.833; ASE = 0.092; 95%CI 0.652–1.00; $p < 0.0001$; almost perfect agreement. Important results are highlighted in bold.

[a] The analysis included 22 assays including commercial vaccines of unknown potency.

2.7 Potency and efficacy: relationship between the antibody titer induced in guinea pigs, bovines and protection against experimental infection in calves

Challenge experiments were conducted in the NBS2 facilities at the National Institute of Agricultural Technology (INTA), as previously described (Romera et al. 2014). Briefly, seronegative calves vaccinated with two vaccines classified by the guinea pig model and bovine calves as having "very satisfactory" or "satisfactory" potency were evaluated. Animals receiving placebo were used as controls and represented an "unsatisfactory" vaccine. Groups of six calves were vaccinated with two doses of each vaccine, 30 days apart. At 90 dpv, animals were challenged by the intranasal route with BoHV-1 virus Los Angeles at a concentration of $10^{7.5}$ $TCID_{50}$/ml. Calves were monitored for infection and disease development. Virus infection was measured by the duration and peak titer of virus shed in nasal swabs and also represented by the "area under the curve of infection" (AUCi) obtained by plotting the infectious titer of virus shed for 14 days after challenge. Disease was measured by the duration and the severity of clinical signs of IBR and also represented by the "area under the curve of clinical signs" (AUCs) obtained by plotting the presence of clinical signs for 14 days after challenge. Both AUCs were calculated using MedCalc® version 11.1.1.0 SAS.

In the challenge experiment (Table 2.9; Figure 2.4), all placebo animals became infected, shedding high titers of virus in the nasal secretions (mean peak BoHV-1 titer: $1 \times 10^{6.9}$ $TCID_{50}$/ml) for 7 days after virus challenge. These animals presented an AUCi of 34.7 and developed typical signs of IBR, characterized by severe bilateral rhinitis (AUCs: 33.8). In contrast, all vaccinated animals challenged at 90 dpv were protected, as they shed significantly less virus and developed a significantly less severe disease compared with the placebo group. Animals vaccinated with vaccine 1 showed significantly higher protection against IBR infection and disease than those vaccinated with vaccine 2, based on all the variables measured (Table 2.9). Significantly different protection rates observed in vaccine 1 and 2 animals were associated with the significantly different mean ELISA Ab titers of each group at 60 dpv.

Vaccines classified as very satisfactory or satisfactory by either ELISA or VN comply with the requirements established by the American 9 CFR USDA and the OIE Manual of Diagnostic Tests and Vaccines of Terrestrial Animals for approval. Regarding protection against infection, a reduction of 1/100 or higher of the titer of infectious virus shed by vaccinated animals as compared with the titer shed by unvaccinated controls is required. In the challenge assay performed with representative vaccines of the very satisfactory and satisfactory categories, in animals vaccinated with both vaccines, the amount of virus shed was significantly reduced when compared with control. Furthermore, virus shed by animals vaccinated with a very satisfactory vaccine was significantly lower than virus shed by the group receiving the satisfactory vaccine. In relation to the duration of clinical signs, the OIE demands a reduction of at least 3 days with respect to the duration of the disease in controls. This requirement was only fulfilled by the very satisfactory vaccine. However, when more appropriate measurements are used to assess the disease, such as the AUC, which considers severity and duration

Table 2.9 Protection against BoH-1 Challenge of Calves Vaccinated with Gold Standard Vaccines of Different Quality

		Bovines							Guinea pig
			Virus shedding			Clinical signs			
Vaccine[a]	***n***	**ELISA Ab titer at 60 pvd**	**Peak of infectious virus titer**	**Duration (days)**	**AUCi**[b]	**Rhinitis severity**	**Duration (days)**	**AUCs**[c]	**Guinea pig ELISA Ab titer 30 pvd**
1	6	$3.37^{A,d}$	2.4^{C}	2.3^{B}	6.2^{C}	1.5^{B}	11^{A}	15.7^{A}	3.1^{A}
2	6	2.68^{B}	5.0^{B}	6.7^{A}	22.2^{B}	1.9^{AB}	12^{A}	20.6^{A}	2.7^{A}
Placebo	9	0.30^{C}	6.9^{A}	7.1^{A}	34.7^{A}	2.5^{A}	14^{A}	33.8^{B}	0.3^{B}

[a] Vaccines previously classified by the guinea pig model and the host species as "very satisfactory" (1) and "satisfactory" (2).

[b] AUCi: area under the curve obtained by plotting the virus titer shed during 14 days after challenge.

[c] AUCs: area under the curve obtained by plotting the severity of the disease registered during 14 days after challenge.

[d] Means in the same column with different uppercase letters indicate significant differences as determined by one-way ANOVA ($p < 0.05$).

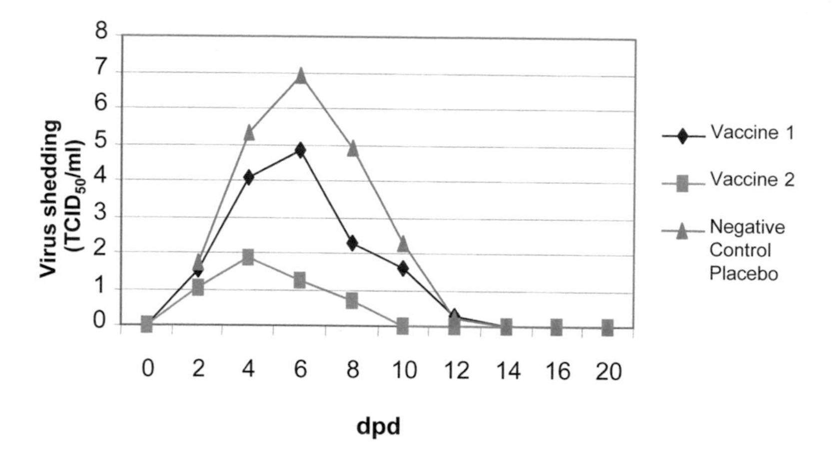

Figure 2.4 Area under the curve of virus shedding of group of calves vaccinated with two vaccines of very satisfactory (A) or satisfactory (B) immunogenicity to IBR.

of clinical symptoms, both vaccine categories significantly reduce the signs of the disease.

The guinea pig model succeeded in adequately predicting not only vaccine immunogenicity but also the efficacy grade when experimentally challenged in bovines. The proposed test does not need complex technology or infrastructure, just an animal facility with guinea pigs and common serological techniques (ELISA, VN) routinely used in virology laboratories, appropriately harmonized with international norms (9CFR 113.216 2014; OIE 2021) and preferably validated following ISO-IEC 17025 norms (Parreño et al. 2010a, b).

2.8 Summary and discussion

Potency is defined as the relative activity of a biological product as determined by a test method conducted on the final product as described and approved by the regulatory agencies. Potency testing of vaccine batches is an important component of the control tests conducted on the final product to confirm consistency of manufacturing and to ensure batch-to-batch quality (Akkermans et al. 2020; Halder et al. 2002; Hendriksen 1999; Taffs 2001).

Specifically, for BoHV-1 killed vaccines, international regulatory policies recommend evaluating vaccine quality in a potency test conducted in seronegative bovine calves. Briefly, the first batch of virus inactivated BoHV-1 vaccine should be tested in five calves susceptible to IBR infection (negative for neutralizing antibodies to BoHV-1). Calves are vaccinated following the manufacturer's recommendations (typically two doses, 30 days apart), while three animals should be left as non-vaccinated controls. Fourteen days after the second dose of vaccine, at least four out of five vaccinated calves should develop virus neutralizing Ab titers of 1:8 or higher. Vaccines not reaching this potency requirement should undergo a challenge test by the intranasal route. Two out of three control calves should develop hyperthermia and clinical signs of IBR, while only one of the vaccinated calves may develop clinical signs; otherwise, the vaccine batch is considered of poor quality. After experimental challenge, the OIE also requires the vaccine to reduce nasal virus shedding at least 100 times and virus excretion period by 3 days compared with control calves (9CFR 113.216 2014; OIE 2021). Thus, for release to the market, a BoHV-1 vaccine must prevent the development of severe clinical signs and markedly reduce virus shedding after experimental challenge. Tests conducted in calves are too expensive and time consuming to be used as a routine test for vaccine batch release. In addition, in many countries where BoHV-1 and 5 infections are endemic, the finding of seronegative bovines from BoHV-1-free herds is becoming increasingly difficult.

The use of a serologic test in guinea pigs, naturally seronegative for BoHV-1 and 5, represents a convenient alternative method to evaluate batch-to-batch vaccine potency, which is in alignment with the 3R initiative of refining, reducing, and replacing animal experimentation (Jungbäck 2011).The BoHV-1 Ab response in guinea pigs and bovines showed a dose-dependent relation to the antigen dose. Under the conditions of the test, the minimum dose of antigen capable of inducing detectable Ab response (detection limit or minimum immunogenicity dose) was in the order of 10^6 $TCID_{50}$/dose of inactivated BoHV-1 in water-in-oil formulation, in both species, as measured by VN and ELISA. The quadratic term of the mathematical equation obtained by linear regression analysis reflected the kinetics of antibody response, indicating that above a certain Ab titer, the Ab response cannot be improved by the addition of higher doses of antigen. The guinea pig model was able to significantly discriminate between vaccines formulated with Ag concentrations differing by 1 $\log_{10}$ and showed very good repeatability and reproducibility, as recommended by the international guidelines (9CFR 113.216 2014; OIE 2021; Taffs 2001; Wright 1999).

The experimental design of the trials conducted in bovine calves attempted to include all the possible sources of variation that can be found in the field (seronegative animals from BoHV-1-free or infected herds and seropositive animals from endemic herds). Despite this high variation, the precision of the target species for testing vaccines containing 10^7 and 10^8 $TCID_{50}$/dose showed CV lower than 20%, indicating that the repeatability and reproducibility of vaccine classification in the natural host by both techniques, VN and ELISA, were quite acceptable.

The sampling time points for comparison between the guinea pig model and the target species were 30 and 60 pvd, respectively, representing the peak or plateau of BoHV-1 Ab response to vaccination with a BoHV-1 inactivated vaccine and the earliest time point of high discrimination capacity.

The determination of split points to generate classification criteria is one of the steps in the development of a laboratory animal model as well as the verification of its predictive ability on the target species. The agreement in the classification of vaccine potency between the two animal species (the lab animal model and the natural host) – based on these split points – was excellent for ELISA and very good for VN, indicating that the developed guinea pig model represents a very good tool to predict vaccine immunogenicity (Parreño et al. 2010a).

In the concordance analysis conducted to evaluate the agreement between the laboratory animal model and the host species for vaccine classification, there are different degrees of misclassification affecting vaccine manufacturers or farmers. In this regard, the use of the two techniques (ELISA and VN) for vaccine classification is recommended as an optimal method to reduce classification discrepancies between the model and the target species. For example, in the first concordance analysis, the classification of two vaccines by ELISA as "unsatisfactory" by the guinea pig model while they were classified as "satisfactory" by the natural host may constitute an economic harm for the vaccine manufacturer if only one technique is considered. One of these cases corresponded to a calibration vaccine containing 10^6 $TCID_{50}$/dose of BoHV-1, misclassified by guinea pig by ELISA and properly classified by the model and the host species by VN. The other was a non-vaccinated control group with pre-existing ELISA antibodies. In this case, the guinea pig properly classified the group as "unsatisfactory" by both techniques while bovine misclassified it by ELISA. On the other hand, when analyzing the agreement between species by VN, there was one vaccine that was classified as "unsatisfactory" by guinea pig but "very satisfactory" by bovine calves. In this case, a failure in the guinea pig immunization was verified, since the same vaccine re-tested 12 months later was properly classified as "satisfactory" by both species and both techniques. On the basis of this incident, it was decided to include a reference vaccine in every immunization test as a positive control in every guinea pig immunization for full validation and transference of the model (Parreño et al. 2010a).

The classification of a vaccine as "satisfactory" or "very satisfactory" in bovines when it is actually unsatisfactory and fails to protect cattle represents a risk for the livestock producers. This kind of misclassification occurred in two cases only by VN, and it was due to a technical error (a laboratory mistake in a given VN run). Again, for these special cases, the use of both techniques (ELISA and VN) allows the source of discrepancy to be clarified. Other kinds of misclassification do not mean an important harm for the industry or the vaccine producer, since the quality of the vaccines classified within both "satisfactory" and "very satisfactory" categories is accepted by the regulatory agencies (9CFR 113.216 2014; OIE 2021). Regarding the second concordance analysis, including commercial vaccines, only 5 out of 22 were classified as "unsatisfactory" by both species and both techniques, suggesting that some vaccine manufacturers should improve the quality of their products in order to reach the proposed standards.

Vaccines assigned to different immunogenicity categories by the proposed statistical model, in guinea pigs and calves, also showed significant differences in protection against infection and disease after challenge in the natural host. Despite their differential immunogenicity and efficacy, both the "very satisfactory" and "satisfactory" vaccines were able to pass the OIE requirements for approval and release. Regarding the evaluation of vaccine efficacy, it is important to highlight that among the different outputs measured after experimental challenge, the AUCi was the best indicator to evaluate protection against challenge and also agreed with vaccine categories in terms of immunogenicity or potency.

Although significant progress has been made in using ***in vitro*** tests to evaluate antigen quality parameters, models to measure veterinary vaccine potency are still based on immunization/challenge assays in the natural host or laboratory animals. The use of ***in vivo*** models for vaccine potency testing and lot release is irreplaceable for the moment (Halder et al. 2002; Hendriksen 1999; Jungbäck 2011; Taffs 2001). From the results obtained, we concluded that the developed guinea pig model is a reliable tool to estimate batch-to-batch vaccine potency, avoiding the problem of finding BoHV-1-seronegative bovines in endemic countries, reducing the variability that can be found in bovines, and significantly reducing the cost and the duration of the test. The model is also ethically compliant with the 3R principle in the use of animals and can also take part in the new consistency approach (Halder et al. 2002; Hendriksen 1999).

The serology results in guinea pigs were statistically validated as a reliable indicator to predict vaccine immunogenicity and protection against challenge in the natural host. Based on these results, the model was transferred to the Argentinean National Health Authorities (SENASA). The use of the model to test IBR vaccines started in 2008 and became official in 2013 (Sanitary Resolution 598.12. SENASA. Argentina).

A laboratory animal model for veterinary viral vaccine potency testing such as the one proposed here would be of use to other regional and international regulatory agencies. In particular, the proposed model is innovative, since it was designed to evaluate killed combined vaccines, and after full validation, it will allow the determination in one immunization assay, using five guinea pigs, of the immunogenicity for all the viral antigens included in the vaccines currently marketed for cattle. In this regard, the Argentinean National Animal Health service has started to test the use of the developed model for IBR vaccine potency testing to control the vaccines present in the local market since 2013, obtaining a significant improvement of vaccine quality since its implementation (see Chapter 6).

Chapter 2 Annex I

Potency testing for IBR in guinea pigs

Guidelines elaborated by PROSAIA biologic ad hoc group (www.prosaia.org/biologicos-veterianrios/) and approved as an American Committee for Veterinary Medicines (CAMEVET) harmonized document in 2015 (https://rr-americas.woah.org/en/projects/camevet/harmonized-documents/?se=IBR&page-nb=1)

Guinea pig conditions. Animals must be more than 30 days old and weigh 400 grams ± 50 grams. For each batch under control, at least SIX (6) GUINEA PIGS will be vaccinated. Males and females may be used, but each group must contain animals of the same sex.

Animal quarantine. Animals will have an adaptation period of SEVEN (7) days, at least, after entering the inoculation room.

Vaccine inoculation. A vaccine of 1/5 the volume of the bovine dose is applied subcutaneously.

Blood extraction to obtain serum samples. It can be collected by cardiac puncture, jugular vein, or auricular vein, without anticoagulant. Samples are clarified by centrifugation, fractioned in 500 μl aliquots, and stored at –20C until analysis. Sample identification: label with date, protocol number, vaccine ID, guinea pig number, DPV time.

Serological controls for potency test in guinea pigs

ELISA assay for the detection of antibodies against IBR

A validated indirect ELISA is used to detect anti-BoHV-1 antibodies. Briefly, plates are sensitized with BoHV-1 virus obtained from infected MDBK cells (positive well) or MDBK cells as negative control of infection (negative well). The optical density of virus and uninfected cells to sensitize plates is determined by crossed titration for each batch produced and is constant for all the plates. Sera are assayed in both wells (+ and –) in six serial fourfold dilutions starting from a minimum dilution of 1/40. The assay is developed using an anti-IgG Ab (H+L) from guinea pig marked with peroxide as detection antibody. H_2O_2/ABTS is used as chromogen substrate system, and reading is performed by an ELISA reader at 405 nm. Currently, the kit is available in commercial form (Figure 2.5).

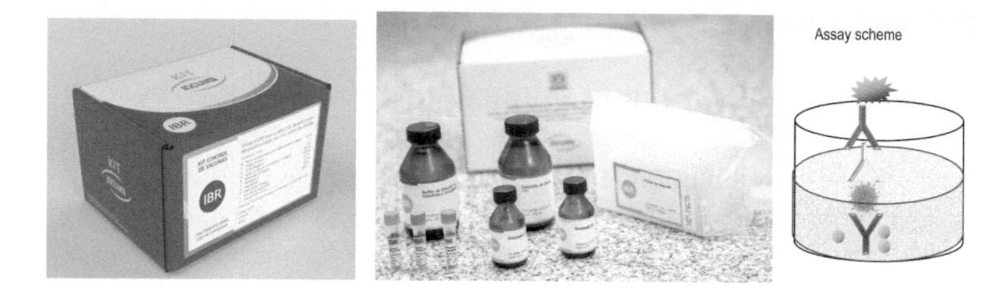

Figure 2.5 ELISA Kit Vaccine quality control IBR (CE-2021-40592704-APN-DLA#SENASA. EX-2021-29837443- -APN-DLA#SENASA).

Reagents

Plate sensitization buffer (carbonate/bicarbonate) pH 9.6

Na_2CO_3 0.159 g.

$NaHCO_3$ 0.293 g. Distilled water q.s.100 ml.

Adjust pH with NaOH/HCl Store at 4C (1–8C).

Citric acid buffer pH 5.0

Citric acid monohydrate 0.960 g.

NaOH 1 N approx.10 ml to reach pH 5.0.

Distilled water q.s. 100 ml.

Adjust pH to 5.0 $\pm$ 0.5 with NaOH or HCl 1 N. Store at 4C (1–8C).

ABTS mother solution

ABTS 0.22 g.

Citric acid buffer 10 ml.

Aliquot 1 ml in plastic tubes.

Store at –20 $\pm$ 5C.

Revealing solution ABTS

ABTS mother solution 300 μl.

Citric acid buffer pH 5 10 ml.

Hydrogen peroxide 30 volume (H_2O_2) 10 μl.

Stop solution
SDS (sodium dodecyl sulfate/sodium lauryl sulfate) in 5% water. Store at room temperature.

Wash buffer (PBS, pH 7.4 – Tween20 0.05%)
PBS pH 7.4 1000 ml.

Tween20 500 µl.

Blocking buffer and diluent (PBS/Tween20 0.05%/OVA 1%, pH 7.4)
Tween20 500 µl.

PBS 1× 1000 ml.

Ovalbumin 10 g.

Aliquot 50 ml in plastic tubes. Store at –20 ± 5C.

Reagents for sensitization
***Positive capture control – Antigen*:** preparation based on MDBK cell cultures infected with BoHV-1 reference strain.

***Negative capture control*:** preparation based on MDBK cell cultures.

***Conjugated detection antibody*:** anti-IgG conjugate marked with peroxidase. The following can be used: affinity purified goat anti-guinea pig IgG (H+L) peroxidase labeled, KPL, cat.# 14-17-06; Peroxidase-conjugated affiniPure goat anti-guinea pig IgG (H+L), Jackson, cat. # 106-035-003. Note: conjugates from other suppliers or produced in-house previously verified/validated as optimal for ELISA assay can be used after corresponding titration with reference sera.

Controls
***Guinea pig positive control*:** serum pool of five guinea pigs vaccinated with two doses of vaccine containing 107 TCID50/ml of BoHV-1 in oil adjuvant (Reference Vaccine). Assays will be accepted when positive controls fall within the mean value ± 1 standard deviation (SD).

Mean value of corrected absorbance ± 1 SD = $0.520 < 0.740 < 0.960$

The given serum, analyzed by seroneutralization, must show an anti-BoHV-1 neutralizing antibodies titer between 2.4 and 3.0 (titer expressed using the Reed and Muench method).

Guinea pig negative control: guinea pig serum whose corrected absorbance in the dilution used is lower than the cut-off of the technique (40% corrected absorbance of positive control).

Reagent blank: PBS. For each control, four wells are used (two positive captures and their two corresponding negative captures). Also, it is advisable to include in each assay a positive sample of known titer and another negative sample.

Dilution and placing of controls

ELISA kit will be provided with standardized controls prepared in the following way. Place 200 μl of diluent (PBS, Tween20, Ova 10%, pH 7.4) in two tubes and 4 μl of positive control serum in one of the tubes and 4 μl of negative control serum in the other. Homogenize.

Add 50 μl of the positive control dilution to wells A7, B7, A8, and B8, 50 μl of negative control dilution to wells A9, B9, A10, and B10, and 50 μl of PBS, Tween20, Ova 10%, pH 7.4 to wells A11, B11, A12, and B12 (reagent blank). Incubate in humid chamber for 1 h at 37C.

Discard contents of the plate. Wash four times. Dry.

To a tube containing 5000 μl of diluent, add the corresponding quantity of conjugated antibody following the dilution used. Add 50 μl per well to the entire plate. Incubate in humid chamber for 1 h at 37C.

Discard the contents of the plate. Wash five times. Dry.

Development, reading, and interpretation

Prepare 5 ml of developing solution. Add 50 μl of developing solution to each well and wait between 10 and 15 minutes with the plate in darkness. Read the assay at 405 nm to ensure that the positive control reaches the optical density expected in the established time range.

Stop reaction by adding 50 μl of stop solution (SDS 5%) to all plate wells and read.

Transfer reading data to a calculation sheet.

Subtract the absorbance of each negative capture from the corresponding positive capture. (Example: H1 minus G1 = ODc [corrected optical densities]).

Calculate the mean of ODc of the positive control (100% PP). Calculate the PP% of each sample in each dilution (PP = [ODc sample/ODc Positive Control)*100]).

Calculate the mean of the replicates of the negative control and its PP%.

Assay acceptance (conformity)

The following described criteria will be applied individually to each plate.

- An assay plate is accepted when the ODc of the positive control is within the established range: 0.520–0.960; the negative control and reagent blank show PP% lower than assay cut-off (40%PP); the positive reference serum titer results in the expected value ± a fourfold dilution (error of the method).
- The antibody titer of a sample is established as the reciprocal of the maximum dilution whose PP% is higher than or the same as the assay cut-off (40%PP).

Reporting results of immunogenic quality of IBR vaccines tested in guinea pigs and sera evaluated by ELISA

Results can be interpreted and used to classify a vaccine in the guinea pig model by ELISA only if the ELISA assay has been accepted and you include a minimum of five animals with results to estimate the mean Ab titer induced by the vaccine.

To validate the assay, sera of guinea pigs immunized with the reference vaccine must result in a mean titer within the established range determined by a control chart that shows the mean value ± 2 standard deviations obtained from a minimum of five tests.

For the vaccine under evaluation, the mean anti-BoHV-1 antibody titer detected by ELISA as the average of the titers of five animals is recorded (log10 of the reciprocal of the maximum dilution whose percentage of positivity is higher than or the same as the assay cut-off, established as higher than or the same as 40% of the positive control). Negative samples in the minimum serum dilution assayed (1/40) are expressed as an arbitrary titer of 0.3 for calculation purposes.

Viral neutralization assay to determine Ab for IBR

Reagents

Virus used: BoHV-1 Los Angeles (LA) reference strain or native BoHV-1 virus strain. Diluted so as to contain 100 tissue culture infectious dose 50% (TCID50).

Controls

Guinea pig positive control: serum pool of five guinea pigs vaccinated with two doses of vaccine formulated with 107 TCID50/ml of BoHV-1 in oil adjuvant (Reference Vaccine) with a neutralizing Ab titer of 2.4–3.0.

Guinea pig negative control: normal guinea pig serum pool (pre-immunized guinea pig sera or unvaccinated controls).

Positive standard: immunized guinea pig serum, with known VN Ab titer.

Negative standard: Normal guinea pig serum.

Cell suspension: a cell suspension of MDBK line containing 200,000–250,000 cells/ml is used.

Sample inactivation: before being used in a VN assay, serum samples, including assay controls, must be heated in a water bath at 56 ± 3C, for 30 ± 5 minutes, to inactivate the complement.

Preparation of working medium: MEM-E supplemented with 1% antibiotic solution (0.5% gentamicin sulfate, 0.7% streptomycin sulfate, 0.2% penicillin G sodium) and 2% bovine fetal serum (BFS).

Viral neutralization assay procedure with fixed virus – variable serum

1. 96-well culture plates are used. Place 75 μl/well of medium in all plates to be used.
2. Design of plate for serums tested: add 25 μl of the sample being tested, in quadruplicate. Start with a minimum dilution of 1/4, which summed to the volume of virus and cells, results in an initial dilution of 1/8. Include standards of known titer at random among the samples to be analyzed.
3. Design of control plate: positive and negative control sera are placed in the same way as the samples. To perform the cell control, add 150 μl of working medium per quadruplicate in 4 rows (16 wells in total). For the control of the 100 TCID50 of virus, three 10-fold dilutions are carried out based on the working dilution. 75 μl of the fourfold dilutions prepared are placed in quadruplicate (pure, 1/10, 1/100, and 1/1000) and 75 μl of medium is added.
4. Carry out fourfold serial dilutions, transferring 25 μl, for all samples and control sera.
5. A toxicity control is carried out for each sample, adding 75 μl of medium in another plate.
6. Prepare the dilution of the work virus (100 TCID50) in the working medium. Add 75 μl of the working dilution of virus to all plates, except for the plate of toxicity controls, the cell control, and the 100 TCID50 control.
7. Carry out three 10-fold dilutions of the working virus (pure, 1/10, 1/100, and 1/1000) and place four replicates of each dilution in a separate plate.
8. Incubate the plates (serum–virus blend) for 1 hour at 37C in an atmosphere containing 5% CO_2.
9. Add 100 μl of the cell suspension containing 200,000–250,000 cells/ml per well to the serum–virus blend in all plates. Incubate plates at 37 ± 1C in an atmosphere containing 5% ± 1% CO_2 for 48–72 h.

Reading and interpretation

After 48–72 h, reading is performed by inspection of monolayers in an optical microscope. Reading is by observation of viral cytopathic effect (CPE) typical of bovine herpes virus. Wells presenting a CPE typical of BoHV-1 are considered positive. In toxicity controls, the monolayer must be observed to be the same as the cell controls, free of CPE, and free of toxic effects. The neutralizing titer of the analyzed serum is obtained by the quantity of protected replicates in the serial dilutions based on the Reed and Muench interpolation method. If a certain serum presents toxicity in the analyzed dilutions, the titer of neutralizing antibodies won't be determined by this technique.

Assay acceptance (conformity)

The assay is accepted when:

- Monolayers of cell controls are in good condition (confluent monolayers, light-refracting cells, with no morphological alterations, with no signs of contamination, and with no BoHV-1 CPE).
- Viral suspension titer contains 100 TCID50, with an admitted range of 50–200 TCID50.
- Positive control shows the expected titer ± 1 well.
- Negative control results are negative. An arbitrary value of 0.3 is assigned for calculation purposes.

Reporting results of immunogenetic quality of IBR vaccines tried/ tested in guinea pigs and sera evaluated by VN

Results can be interpreted and used to classify a vaccine in the guinea pig model by VN only if the viral neutralization assay has been accepted and you include a minimum of five animals with results to obtain the mean Ab titer induced by the vaccine. To validate the assay, sera of guinea pigs immunized with the reference vaccine must show a mean titer within the range established by a control chart, which shows the mean value ± 2 standard deviations obtained from a minimum of five samples.

For the vaccine under evaluation, mean neutralizing anti-BoHV-1 Ab titer obtained by the Reed and Muench method of the five immunized guinea pigs must be reported. Negative samples in the minimum serum dilution assayed (1/8) are expressed as an arbitrary titer of 0.3 for calculation purposes.

Validation criteria for guinea pig testing

Potency testing in guinea pigs is considered valid when the mean Ab titer obtained from animals vaccinated with a reference vaccine is the expected value (2.4), and unvaccinated control animals (controls) remain seronegative for Ab against BoHV-1 throughout the experience.

Vaccine approval criteria by potency testing in guinea pigs

a) ELISA

All sera of animals immunized with a control vaccine will be evaluated. FIVE (5) sera with the highest titers obtained will be selected, and an average will be calculated on that basis. For the APPROVAL of the vaccine submitted to control, mean Ab titers at 30 dpv must be higher than or the same as 1.93 for the ELISA technique for BoHV-1 (see Table 2.3).

b) Viral neutralization

All sera of guinea pigs immunized with the vaccine submitted to control will be evaluated. FIVE (5) sera with the highest titers obtained will be selected, and an average will be calculated on that basis. For APPROVAL of the vaccine submitted to control, mean Ab titers at 30 dpv must be higher than or the same as 1.31 for the VN technique for BoHV-1 (see Table 2.5).

Harmonization of assays for the region

A positive and negative control serum panel and reference vaccines will be elaborated and made available for regional users upon request to harmonize the results obtained for each assay laboratory adopting the control method. Local reference sera (from guinea pigs and bovines) will be traceable in the described techniques to European reference bovine sera (EU1, EU2, and EU3) provided by OIE Reference Laboratories.

3
Guinea Pig Model to Test the Potency of Rotavirus Vaccines

Development and statistical validation of a guinea pig model and its associated serology assays as an alternative method for potency testing of bovine rotavirus vaccines.

3.1 Introduction

Neonatal calf diarrhea (NCD) is a common disease affecting newborn calves in both dairy and beef herds worldwide. It is a multifactorial disease that affects newborn calves up to 3 months of age. The most critical period is the first few weeks of life, because the prevalence of NCD is higher between 1 and 21 days of age, with a peak incidence at 2 weeks, but it can extend up to 30–45 days of age (Bartels et al. 2010; Bertoni et al. 2021). Affected calves show watery feces of increased frequency. The loss of liquids through diarrhea induces progressive dehydration and acidosis. If the disease is allowed to progress untreated, dehydration will cause the loss of electrolytes (sodium, potassium, chloride, and bicarbonate). Calves develop metabolic acidosis, which results in rapid death (in less than 2 days) if not treated promptly (Trefz et al. 2012).

The causes of NCD can be infectious or non-infectious. However, infectious diarrheas are the ones that cause the greatest mortality problems (Blanchard 2012). The etiology of this disease involves viral, bacterial, and parasitic agents, including rotavirus group A (RVA), coronavirus, pathogenic *Escherichia coli* strains with virulent factors, *Salmonella* spp., *Clostridium perfringens*, *Cryptosporidium parvum*, and Coccidia (Foster and Smith 2009).

Within this complex etiology, bovine RVA infection is considered the most important viral agent causing diarrhea in calves worldwide (Bertoni et al. 2021; Cho et al. 2013; Cho and Yoon 2014). Actually, rotavirus was one of the first identified viral causes of diarrhea in calves. The virus typically affects calves younger than 3 weeks old, but they may be up to 8 weeks of age. The incubation period is approximately 24 h; diarrhea can be mild to severe, but it usually resolves within 2–5 days. Calves become infected

DOI: 10.1201/9781003373148-3

after ingesting the virus from fecal contamination of the environment. The virions are highly resistant to chemical and physical inactivation, remaining infectious for up to 4 months in humid environments (10-4°C) or in frozen feces. Infected animals shed around 10^{11} RVA particles per gram of feces. Rotaviruses replicate in the mature enterocytes of the small intestine. The tips of the villi undergo cell lysis, resulting in villous blunting and ischemia, causing maldigestive, malabsorptive, and secretory diarrhea. By 3 months of age, calves are generally no longer susceptible to infection by this virus (Blanchard 2012; Foster and Smith 2009).

Like influenza viruses, RVA has a dual serotype specificity based on the two outer surface proteins that induce neutralizing antibodies: VP7 and VP4. The VP7 glycoprotein is prominent; it makes up the surface of the external capsid, and its variants are called G-types; and the viral spike VP4 protein is sensitive to protease and cleavage by trypsin in two proteins, VP8* and VP5*. The variability of the tip of the protein represented by the VP8* domain is the origin of the P-types (Matthijnssens et al. 2011).

Among the RVA circulating in cattle, the G-types have been reported to be G1, G2, G3, G4, G5, G6, G8, G10, and G15, and the P-types P[1], P[5], P[11], P[14, P[17], and P[21]. However, only G6, G10, and G8 associated with P[5], P[11], and P[1] are considered epidemiologically important (Brunauer, Roch, and Conrady 2021; Papp et al. 2013).

The combination G6P[5] is the prevalent strain circulating in beef herds, while G10 and G6 associated with P[11] are found in dairy herds (Badaracco et al. 2013; Badaracco et al. 2012; Bertoni et al. 2020; Bertoni et al. 2021). Viruses with distinct G- and P-types do not cross-protect. Thus, it is important to assess the characterization of the rotavirus circulating in a region/country in order to evaluate whether the vaccines need to be updated.

The prevention of the disease is based on passive immunity strategies. Cows are vaccinated during the late stage of pregnancy (ideally, 60 and 30 days pre-calving) with RVA inactivated vaccines to increase the maternal passive immunity transferred to their calves via colostrum intake. The killed vaccines used to prevent RVA calf scours are aqueous or oil formulations including the most prevalent RVA variants found in cattle G6P[5] and G10P[11]. Inactivated RVA can be present in oil or aqueous adjuvanted vaccines together with different combinations of the following antigens: *E. coli*, *Salmonella*, *Clostridium*, and coronavirus.

The prevention measures used worldwide to control NCD due to bovine RVA are focused on trying to reduce the severity of the disease in the calf and the titer of infectious virus excreted into the environment. As a strategy to increase the specific passive immunity in neonates, vaccination of pregnant mothers during the last stage of pregnancy is recommended (60 and 30–45 days before calving) to favor the transfer of passive maternal antibodies (Abs) and immune cells to the newborn via colostrum intake during the first hours of life. This passive immunity strategy based on dam vaccination to protect the calves is able to significantly control the disease and reduce, but not eliminate, the virus shedding (Parreño et al. 2004; Saif and Fernandez 1996).

International regulatory bodies (Animal and Plant Health Inspection Service [APHIS], USA; the European Medicines Agency Committee for Veterinary Medicinal Products [EMEA-CVMP], EU; World Organisation for Animal Health [OIE]; International Cooperation on Harmonisation of Technical Requirements for Registration of Veterinary Medicinal Products [VICH]) have not made recommendations for the development and control of rotavirus vaccines. However, given the health impact of this neonatal pathology in livestock farms worldwide, the development and implementation of a potency control test for these vaccines in laboratory animals is considered important.

In the present chapter, we describe the development and statistical validation of the National Institute of Agricultural Technology (INTA) guinea pig model to be used as an alternative method for vaccine potency testing to predict the immunogenicity and efficacy of RVA vaccines in bovines. Guinea pigs, raised under controlled conditions, possess the advantage that they are naturally seronegative for antibodies (Ab) to RVA but develop a strong Ab response after RVA immunization, resulting in an excellent model for vaccine potency testing.

The test allows quality control of the batches of vaccine that are released on the market to check consistency among batches and stability through time. The vaccine potency in guinea pigs has been statistically validated against the potency of the vaccines obtained in the target species. The vaccine classification by the model has an optimal degree of agreement to predict the potency of each batch of vaccine in bovines. Likewise, the potency of vaccines classified according to the guinea pig model was associated with the degree of protection against infection and diarrhea in neonatal calves receiving colostrum from vaccinated cows and further challenged with rotavirus. The proposed model represents a predictive tool for the degree of protection conferred by colostral antibodies against viral discharge in neonatal calves challenged with RVA (Parreño et al. 2004).

This test allows the vaccine producers and the National Veterinary Health Authorities to control the immunogenicity of the RVA vaccines released to the market in a harmonized way. As explained in Chapter 2 for BoHV-1, the validation for rotavirus involved the study of the kinetics of the Ab response in the animal model and the target species, a regression and a classification tree analysis applied to the dose–response curve to define categories for vaccines qualification, a concordance analysis between the laboratory animal model and the natural host, and a confirmation study of vaccine efficacy where neonatal calves were fed with colostrum from vaccinated and non-vaccinated cows and then challenged with rotavirus. The potency of the vaccines in terms of mean Ab titer obtained in guinea pigs was associated with the Ab titer in the vaccinated cows and their calves, and protection of calves against rotavirus infection and disease.

3.2 Dose–response assay: anti-RVA antibody responses and discriminatory power in guinea pigs and bovines

Two sets of four trivalent oil vaccines containing serial 10-fold dilutions of bovine rotavirus (RVA) reference strains UK G6P[5] and B223 G10P[11] together with *E. coli* J5 were formulated in oil adjuvant. The viruses were binary ethylenimine (BEI) inactivated, and the emulsions were made in pilot production under industrialized conditions to represent the oil-adjuvant commercial vaccines present in the market that are administered to the dam in the last stage of pregnancy for the prevention of NCD.

Each vaccine set (1 and 2) consisted of four antigenic concentrations: A1-A2:10^7; B1-B1:10^6; C1-C1:10^5; and D1-D2:10^4 FFU of RVB UK G6P[5] and B223 G10P[11] per dose. A placebo containing all the vaccine components, except the antigens, was used as negative control (Table 3.1 and Figure 3.1).

Table 3.1 Dose–Response for Rotavirus Vaccines. Experimental Design

Vaccine set	Bovine RVA UK G6P[5], B223 G10P[11], concentration ($TCID_{50}$/dose)	Beef herd[a]	Dairy herd[a]	Guinea pigs[b]	
		Set 1 + Set 2	Set 1 + Set 2	Set 1	Set 2
A1–A2	10^7	5 + 5	5 + 5	5 + 5	5 + 5
B1–B2	10^6	5 + 5	5 + 5	5 + 5	5 + 5
C1–C2	10^5	5 + 5	5 + 5	5 + 5	5 + 5
D1–D2	10^4	5 + 5	5 + 5	5 + 5	5 + 5
Placebo	Adjuvant media	5 + 5	5 + 5	5 + 5	5 + 5
Total		50	50	100	

[a] Each vaccine set was tested in five animals (Set 1 in five animals and set 2 in five animals).

[b] Each vaccine set (1 and 2) was tested in two independent assays in guinea pigs.

Figure 3.1 Dose–response study for rotavirus vaccines. Beef and dairy cows and heifers received two doses of 3 ml of oil-adjuvanted vaccine 30 days apart. A total of 100 bovines and 100 guinea pigs were enrolled in the study for the statistical validation of the lab animal model.

These two sets of reference vaccines were tested in heifers belonging to a beef herd and cows from a dairy herd (I = 5 per group) and in two independent experiments in guinea pigs (Figure 3.1). Regarding the immunization of bovines, it is important to note that rotavirus infection is endemic and that all calves are infected and generate antibodies against the agent during their first months of life. This phenomenon makes it impossible to find seronegative cattle in which to evaluate the potency of these vaccines without interference. Based on this, to carry out the proposed dose–response study, 6-month-old non-pregnant heifers belonging to a beef herd were selected. Animals of this category were used in order to apply the vaccines to young animals that no longer had passive colostral anti-RVA antibodies and that presented low levels of their own antibodies against RVA exposure early in life (only animals with anti-RVA IgG1 titers ≤4096 were enrolled in the field trial). In this way, the heifers included in the trial represent "naïve" animals that will be prime-vaccinated with the vaccines under study. The heifers were vaccinated with two doses of 3 ml of vaccine, 30 days apart. Serum samples were taken at 0 and 60 post-vaccination days (pvd). A group receiving placebo and a group of non-vaccinated animals were included as negative controls. In addition, to obtain data on the target category and in the dairy herds, the vaccines were administered to pregnant Holstein cows from a dairy herd. In this case, the females were vaccinated 60 and 30 days before calving. Serum samples were taken at the time of each vaccination and serum and colostrum at calving (60 pvd).

In parallel, each set of vaccines was administered to groups of five guinea pigs in two independent assays. Guinea pigs received two doses of vaccine corresponding to one-fifth of the bovine dose (0.6 ml, 21 days apart) (Figure 3.1). The minimum number of experimental units (animals per group; n = 5) was calculated to achieve a statistical power of at least 83% to discriminate between vaccines containing bovine RVA concentrations differing by one $\log_{10}$.

Rotavirus IgG Ab response to vaccination in guinea pigs was measured by enzyme-linked immunosorbent assay (ELISA) and statistically compared with the IgG1 Ab response in vaccinated cows. Both ELISA assays were standardized under ISO 17025 standards; the validation parameters are summarized in Table 3.2. Bovine IgG1 isotype was specifically selected as a marker to evaluate the quality of bovine RVA vaccines, because IgG1 is the specific isotype that is actively transferred and concentrated from serum to colostrum in the bovine mammary gland. IgG1 represents the main isotype of passive maternal antibodies transferred to the neonatal calves after colostrum intake during the first 24 hours of life. The antibody titers were expressed as the inverse of the highest serum dilution with a positive value by ELISA (percentage of positivity higher than the cut-off of the assay). The titers were log-transformed for statistical analysis.

Table 3.2 ELISA Validation Parameter for the Detection of IgG1 Antibodies in Bovine and IgG in Guinea Pigs for Vaccine Potency Testing

ELISA validation parameters	IgG anti-RVA (guinea pigs)	IgG1 anti-RVA (bovines)
Cut-off	11%PP[a]	12%PP[a]
Sensitivity	98.5%	97.7%
Specificity	97.7%	100%
Intermediate precision (3 years)	CV[b] = 24%	CV[b] = 13.5%
Accuracy (ROC[c] analysis)	98.5%	100%

[a] PP = percentage of positivity compared with a positive control run in every plate and every run.

[b] CV = coefficient of variation.

[c] ROC = receiver operating characteristic curve.

As explained for infectious bovine rhinotracheitis (IBR) in Chapter 2, in the case of RVA, the obtained results were used:

i) to establish the optimal sampling time for comparison between the guinea pig model and bovines.
ii) to evaluate the ability of each species to discriminate among vaccines of different antigen concentration (analysis of variance [ANOVA]).
iii) to estimate the split points for vaccine classification by regression and classification tree analysis.

In the dose–response assays, most of the guinea pigs and the beef cattle generated results. However, only 25 out of 50 cows that started the study with Ab titers ≤4096 could be used in the field trial conducted in the dairy farm. In this way, in guinea pigs, four mean Ab titers were obtained for each antigen concentration, and for cattle, the analysis was carried out with only three mean Ab titer values, two means by concentration (A to D) from the beef heifers immunized with vaccine sets 1 and 2, and one mean Ab titer from dairy cows starting the trial with RVA Ab titers ≤4096 and vaccinated with 1 or 2 vaccine sets.

To study the discriminatory capacity, anti-RVA Ab titers in guinea pigs and increases in Ab titers in cattle determined by ELISA were analyzed using an ANOVA model and subsequent comparisons of means by Dirienzo Guzman Casanovas (DGC) test (Infostat manual (InfoStat 2008)). To determine the cut-off points, linear regression and classification tree analysis were used (Lemon et al. 2003).

Both species showed a dose–response according to the RVA antigen concentration in the vaccine. As previously described for IBR, the time points to compare the vaccine potency between the target species and guinea pig were 60 pvd (calving) in vaccinated dams and 30 pvd in guinea pigs. Both species were able to discriminate among vaccines containing 1 or 2 $\log_{10}$ difference in their antigen concentration. When the increment in the Ab titer after vaccination was studied, bovines were able to discriminate between vaccines containing 10^7 from those with 10^6–10^5 and 10^4 tissue culture infective dose $(TCID)_{50}$ RVA/dose. The guinea pig model was able to discriminate among vaccines with 10^6 $TCID_{50}$ or higher and those with 10^5 and 10^4. In both species, vaccines containing 10^4 $TCID_{50}$ did not induce Ab responses to RVA (Table 3.3).

3.3 Selection of quality split points for rotavirus vaccines: classification tree analysis

In bovines, the split points for vaccine classification were estimated in terms of increment of Ab titer after vaccination as well as the final Ab titer reached

Table 3.3 Vaccine Classification Criteria by ELISA

	Bovine			Guinea pig	
Rotavirus vaccine concentration ($TCID_{50}$/dose)	***n***	**Δ[a] anti-RVA IgG1 Ab titer T60–T0**	**Mean Anti-RVA IgG1 Ab at 60 pvd (calving)**	**n**	**Mean anti-RVA IgG Ab titer at 30 pvd**
10^7	15	1.04 A	4.00 A	20	4.79 A
10^6	14	0.63 B	3.52 AB	18	4.67 A
10^5	15	0.44 B	3.23 AB	17	4.00 B
10^4	14	0.10 C	3.25 AB	16	0.70 C
Placebo	17	−0.03 C	2.91 B	20	0.30 C

Means in the same column with different uppercase letters differ significantly. One-way ANOVA. Bonferroni. $p < 0.0001$.

[a] Difference between the initial Ab titer and the final Ab titer after vaccination.

at 60 pvd (calving). Regarding the Ab increment at 60 pvd, a linear quantitative relationship between antigen dose and Ab response (t60–t0 IgG1 Ab titer) was observed ($R^2 = 82\%$, $p < 0.0001$) (Figure 3.2a). From this linear regression curve and using the non-parametric classification tree method, it was estimated that the minimum increase that should be recorded in the anti-RVA IgG1 Ab titer of bovines vaccinated with vaccines containing virus concentrations greater than 10^4 $TCID_{50}$/dose was 0.33, while the minimum increase in anti-RVA IgG1 Ab in cattle immunized with vaccines containing virus concentrations of 10^7 $TCID_{50}$/dose or higher was 0.75. These cut-off points were used to assess the degree of agreement between the target species and the guinea pig model to classify vaccine potency. In addition, when analyzing the final IgG1 Ab titer to RVA, a satisfactory vaccine must induce a final IgG1 Ab titer of 3.7 or higher.

The dose–response curve in guinea pigs followed a second-degree linear quantitative relationship between the Ag concentration per dose and anti-RVA Ab response ($R^2 = 95\%$, $p < 0.0001$). The vaccine quality split points in guinea pigs were estimated as the value that better discriminates the vaccines previously classified by bovines using classification tree analysis (Figure 3.2b). The classification tree analysis of the Ab titers obtained in bovines and guinea pigs allowed vaccine classification into four potency categories (Table 3.4). The animal health authorities in each country and regulatory agencies are able to select the desirable classification to conduct the quality control in their jurisdictions. Vaccines inducing a mean Ab titer lower than 1.96 in guinea pigs and a final IgG1 Ab titer less than 3.53 in bovines were considered not satisfactory, while vaccines inducing mean Ab titers

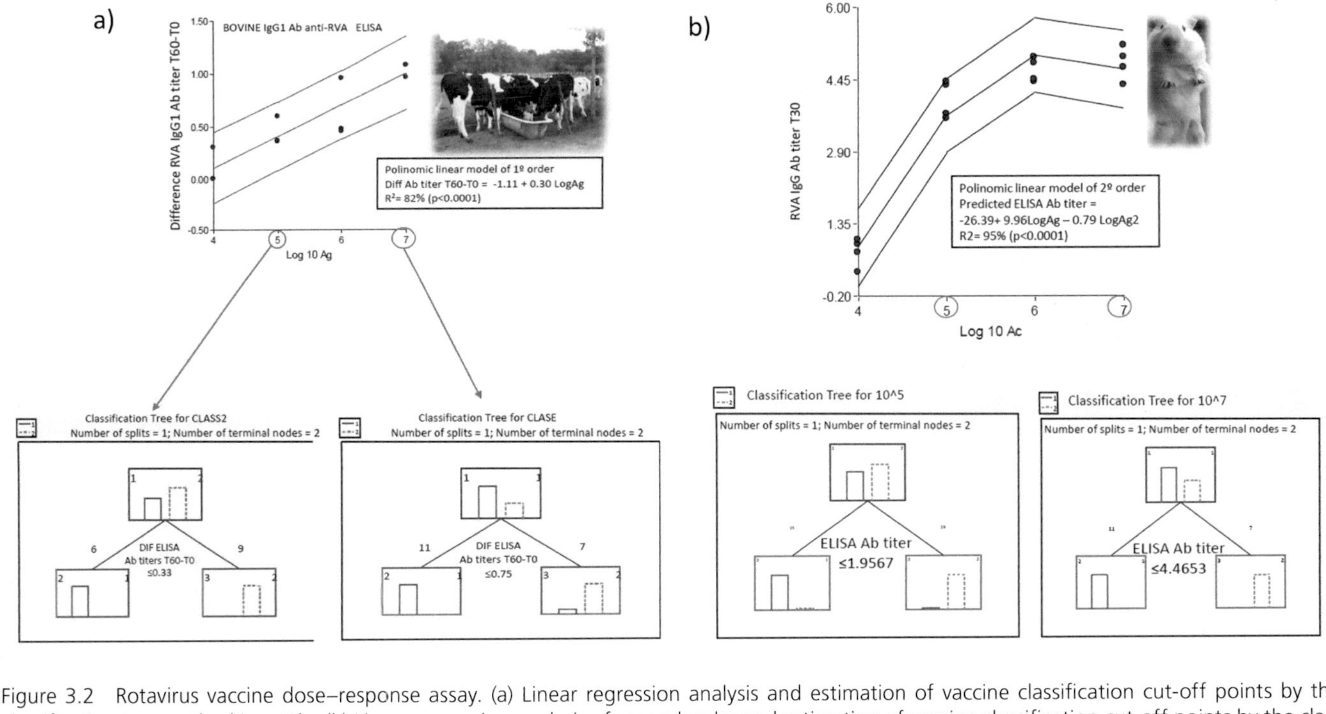

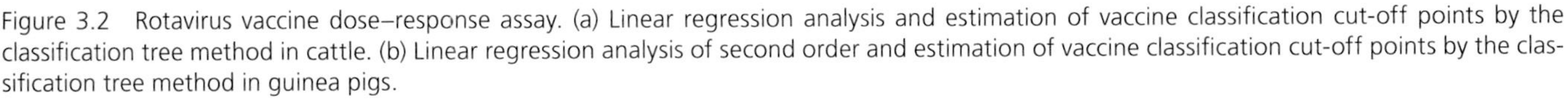

Figure 3.2 Rotavirus vaccine dose–response assay. (a) Linear regression analysis and estimation of vaccine classification cut-off points by the classification tree method in cattle. (b) Linear regression analysis of second order and estimation of vaccine classification cut-off points by the classification tree method in guinea pigs.

greater than 4.00 in guinea pigs and an increase in IgG1 anti-RVA titer in vaccinated cattle of 0.40 with a final IgG1 Ab titer 3.70 at calving (60 days post-vaccination [dpv]) were considered as intermediate or satisfactory, and vaccines inducing Ab higher than 4.75 in guinea pigs and an increment in IgG1 Ab titer in bovines were very satisfactory (Table 3.4).

3.4 Concordance analysis between bovines and guinea pigs for rotavirus vaccine classification

To study the degree of concordance between the guinea pig model and bovines, 43 cases were analyzed by the weighted kappa statistic. The study included the 12 mean Ab titers obtained from the guinea pigs and bovines vaccinated with the calibration vaccines of the dose–response trial (gold standard vaccines), 10 groups that received placebos or were not vaccinated (negative controls), 4 experimental vaccines of known potency (gold standard), and 17 commercial vaccines (Table 3.5).

Concordance analysis using 43 comparison lines, including calibration, gold standard, and commercial vaccines together with placebos and negative controls (Table 3.6), showed almost perfect agreement between the lab animal model and bovines for the classification of RVA vaccines (weighted kappa: 0.825; $p < 0.0001$) (Table 3.6).

From the kappa value obtained, it is concluded that the concordance between the model and the bovines is considered very good. The 16 rejected lines corresponded to 10 control groups of control animals not vaccinated or vaccinated with placebos and 6 vaccinated groups, of which 3 were vaccinated with commercial vaccines and 3 with gold standard vaccines. The five rejected vaccines corresponded to the two dose–response vaccines formulated with 10^4 $TCID_{50}$/dose of rotavirus tested in two groups of breeding cattle and one group of dairy cattle, two commercial neonatal calf diarrhea (NCD) double oil emulsion vaccines evaluated at 9 months post-production (9 mpe), and two vaccines tested at 24 mpe, corresponding to the expiration date.

When only the commercial vaccines and negative controls were tested, and the reference vaccines were excluded from the analysis, the concordance remained, with substantial agreement (weighted kappa: 0.762, $p < 0.0001$) (Table 3.7).

The lines with discordant classifications corresponding to five vaccines classified as very satisfactory by cattle and satisfactory by the guinea pig model can be explained by the fact that the lab animal model is more rigorous,

Table 3.4 Rotavirus Vaccine Classification Criteria by ELISA in Guinea Pigs and Bovines

	Potency (RVA concentration per dose. $TCID_{50}$/dose)			
Species	**Not satisfactory ($<10^5$)**	**Intermediate (10^5–10^6)**	**Satisfactory (10^6–10^7)**	**Very satisfactory ($>10^7$)**
Guinea pig	$x < 1.96$	$1.96 \leq x < 4.00$	$4.00 \leq x \leq 4.75$	$x \geq 4.75$
Bovine Dif. IgG1 Ab titer t60–t0	$X < 0.33$	$0.33 \leq X < 0.40$	$0.40 \leq x \leq 0.75$	$X \geq 0.75$
Bovine Final IgG1 Ab titer at 60 pvd	$X < 3.53$	$3.53 \leq X < 3.70$	$X \geq 3.70$	

Cut-off points determined as the $\log_{10}$ anti-RVA antibody titer (IgG in guinea pigs) determined by ELISA. In bovines, the increase in post-vaccination anti-RVA IgG1 Ab is analyzed with respect to the basal Ab titer, also determined by ELISA, present in the serum of animals vaccinated with the unknown vaccine. (x) Average Ab titer of groups of five guinea pigs evaluated at 30 days post-vaccination (dpv); (X) groups of five cattle evaluated at 60 dpv. Cattle receive two doses of vaccine with an interval of 30 days and are sampled at 0 and 60 dpv. Guinea pigs receive two doses of vaccine (1/5 bovine dose volume) 21 days apart and are sampled at 0 and 30 dpv.

Table 3.5 Vaccines Included in the Concordance Analysis

Type of vaccine	Vaccine composition (antigens)	RVA concentration ($TCID_{50}$/dose)	Number of vaccines tested	Number of vaccinated bovines	Number of vaccinated guinea pigs	Number of comparative assays[c]
Calibration vaccines for the Dose–response curve[a]	*E. coli* + RVA G6P[5] G10P[11]	1×10^7, 10^6, 10^5, 10^4, placebo	8	90	90	12
Gold standard vaccine of known concentration[b]	G6P[5] monovalent	1×10^7	1	5	5	2
	G10P[11] monovalent	1×10^7	1	5	5	2
	G6P[5]G10P11][c]	1×10^7	1	5	5	2
Commercial vaccines[c]		unknown	14	150	75	14
Placebo/unvaccinated controls			10	50	50	11
Total			**35**	**305**	**230**	**43**

[a] Vaccines detailed in Table 3.1 and Figure 3.1.

[b] Gold standard vaccines: vaccines of known antigen concentration (potency) prepared under industrial conditions.

[c] Commercial vaccines of unknown antigen concentration representing the available vaccines in the local market.

Table 3.6 Concordance between the Guinea Pig Model and Bovines for the Classification of RVA Vaccines Considering Three Categories and Two Split Point

Anti-RVA Ab titers determined by ELISA; *n* = 43			
Bovine / Guinea pig	Not satisfactory (Xdif t60–t0 < 0.33)	Intermediate/ satisfactory (0.33 ≤ x < 0.75)	Very satisfactory (0.75 ≤ x)
Not satisfactory (x < 1.96)	**16**	0	0
Intermediate/ satisfactory (1.95 ≤ x < 4.46)	0	**8**	5
Very satisfactory (4.46 ≤ x)	0	2	**12**

Weighted Kappa 0.825; $p < 0.0001$; almost Perfect agreement.
Sample size 43; 43 comparison lines. Placebo/non-vaccinated control groups = 10; Experimental vaccines of known concentration = 4; Doses-response reference vaccine = 12; Commercial vaccines = 17.

Table 3.7 Concordance between the Guinea Pig Model and Bovines to Classify RVA Vaccines Using Two Split Points and Three Classification Categories

Anti-RVA Ab titers determined by ELISA ; n = 31			
Bovine / Guinea pig	Not satisfactory (Xdif t60–t0 < 0.33)	Satisfactory (0.33 ≤ x < 0.75)	Very satisfactory (0.75 ≤ x)
Not satisfactory (x < 1.96)	**14**	0	3
Satisfactory (1.95 ≤ x < 4.46)	0	**3**	5
Very satisfactory (4.46 ≤ x)	0	2	**7**

Weighted kappa 0.762; $p < 0.0001$; almost perfect agreement

since the animals are true naïve, while the bovines are seropositive with pre-vaccination Ab titers that will be boosted by vaccination, and the Ab response will depend on the degree of virus circulation in each specific herd. On the other hand, the two vaccines approved in guinea pigs and as classified satisfactory in cattle only reached 0.03 and 0.15 points of difference from the cut-off line.

3.5 Potency and efficacy: relationship between the antibody titer induced in guinea pigs, bovines and protection against experimental infection in calves

Colostrum-deprived calves were confined to isolation boxes and were fed 1 liter of colostrum from cows vaccinated with a very satisfactory RVA vaccine or colostrum from cows vaccinated with an intermediate-satisfactory vaccine within the first 6 hours after birth and then fed Ab-free milk (Figure 3.3). Colostrum-deprived calves were assigned as the negative control group. All animals were challenged orally with 10^5 focus-forming units (FFU) of bovine RVA UK G6P[5]. Calves fed colostrum from cows vaccinated with the very satisfactory vaccine showed significantly higher protection against RVA diarrhea than calves fed colostrum from cows receiving the intermediate-satisfactory vaccine. This latter group did not differ from the colostrum-deprived calves. All animals shed virus after the experimental infection. Only calves fed colostrum from cows vaccinated with the very satisfactory vaccine showed a significant delay in the onset of virus shedding compared with the colostrum-deprived calves (Table 3.8).

In relation to the potency of the vaccine in guinea pigs and cattle and its efficacy in preventing diarrhea in calves, the studies carried out indicated that the group of calves receiving 1 liter of colostrum from cows immunized with the vaccine of intermediate quality reached titers of serum Ab between 2.4 and 3.01 and developed severe diarrhea, while calves receiving 1 liter of colostrum from cows vaccinated with very satisfactory vaccines (Ab titer greater than 3.70) reached serum Ab titers between 4.21 and 4.81 and presented a significant reduction in the severity of diarrhea.

3.6 Discussion and conclusions

The obtained results indicate that the validated guinea pig model represents a reliable, economic, and rapid tool for a potency test for rotavirus vaccines, aligned with the refining, reducing, and replacing (3R) principle. The potency obtained by the model represents a predictive value of the vaccine potency in pregnant cows and vaccine efficacy to prevent RVA diarrhea in neonatal calves. Further vaccine testing of different brands and formulations in different countries will improve and make more robust the present statistical validation.

In addition, the model represents an excellent tool for batch-to-batch RVA vaccine potency testing to comply with the consistency standards. It can be also used to test the extension of vaccine expiration dates.

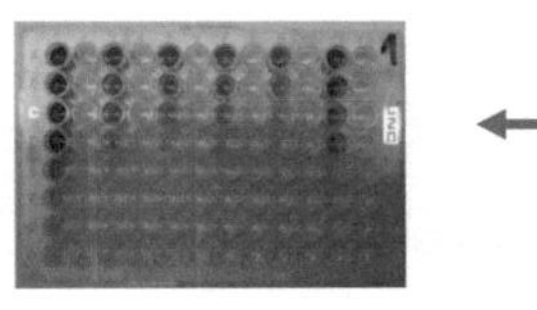

Figure 3.3 Calf model of rotavirus infection and disease. (From Parreño et al. 2004.)

Table 3.8 Result of Vaccine Protective Efficacy in Experimentally Challenged Calves

Treatment	RVA IgG Ab titer in guinea pigs	RVA IgG1 Ab titer in cow at calving	*n calves*	RVA IgG1 Ab titer in cow at calving	Diarrhea				Virus shedding			
					% animals affected	Onset (days)	Duration (days)	Severity	% animals affected	Onset (days)	Duration (days)	Average RVA titer (FFU/ml)
Very satisfactory vaccine	4.94	4.81	6	4.52 A	67%	5.5 A	0.83 A	4 A	100%	3.3 A	8	3.7
Intermediate vaccine	4.36	3.61	9	2.94 B	100%	2.2 B	7.44 B	20 B	100%	2.11 B	8	4.0
Colostrum-deprived calves	Neg.	Neg.	10	Neg.	100%	1.6 B	8.30 B	24 B	100%	1.6 B	6	4.4

Although the ANOVA showed good discrimination between increasing concentrations of antigen, the lack of adjustment of a linear function that relates both variables does not allow reliable cut-off points to be obtained from the regression analysis. The classification tree method, given its greater flexibility as it is a non-parametric technique that does not require specifying a function, allowed an adequate estimation of these points (Lemon et al. 2003).

The model was adopted by the Argentinean animal health service (SENASA) as the official test for RVA for vaccine quality control in Argentina (SENASA sanitary resolution 598.12). The proposed cut-off point 1.96/3.53 (which only rejects very low-quality vaccines) is being used by SENASA to carry out the initial survey of the vaccines on the market, and then, based on the results obtained, the requirement will be increased in order to promote the systematic improvement of vaccine quality. The implementation of the guinea pig model helped to guarantee the presence of effective products in the local market. The guinea pig model for RVA vaccine potency testing was voted by the country members of the American Committee for Veterinary Medicines (CAMEVET) as a harmonized guideline in the meeting held in 2015 (https://rr-americas.woah.org/es/proyectos/camevet/documentos-armonizados/?se=rotAVIRUS&page-nb=1). We hope the application of the guinea pig model will be extended as a harmonized control of rotavirus vaccine quality in the Americas and worldwide, adding this viral agent as a topic in the World Organisation for Animal Health Manual of Diagnostic Tests and Vaccines for Terrestrial Animals.

Chapter 3 Annex II

Potency test for RVA vaccines in the guinea pig model

Guidelines elaborated by PROSAIA biologic ad hoc group (www.prosaia.org/wp-content/uploads/2021/04/Prosaia-3-GB-Gu%C3%ADa-de-Potencia-vacunas-Rotavirus-Español-Abril-2014-enviada-a-Camevet.pd) and approved as a CAMEVET harmonized document in 2015 (https://rr-americas.woah.org/en/projects/camevet/harmonized-documents/?page-nb=1&se=rotavirus)

Process for potency test in guinea pigs

Guinea pig conditions. Animals older than 30 days of age are used, and the weight must be 400 grams ± 50 grams. For each control series, at least SIX (6) GUINEA PIGS will be vaccinated. Males or females may be used, but each group must contain animals of the same sex.

Animal quarantine. The animals are installed in the racks of the experimentation room and remain at least SEVEN (7) days without treatment, for adaptation to the management of feeding, care, and cleaning of the sector, prior to the start of the tests.

Vaccine inoculation. The vaccine is applied parenterally subcutaneously at one-fifth of the volume of the bovine dose.

Blood extraction to obtain serum samples. It can be done by cardiac puncture, jugular vein, or marginal auricular vein.

Sample processing. Obtaining blood without anticoagulant, separation of the clot and obtaining of the blood serum. Clarification by centrifugation. Fractionation in aliquots of 500 μl. Storage at -20°C until analysis. Label: protocol number, vaccine, animal number, and post-vaccination time.

ELISA assay for the detection of antibodies against rotavirus

It is important to note that guinea pigs are a species free of rotavirus group A, and therefore naturally seronegative for Ab against this viral agent, and are only positive after being experimentally immunized with this agent. A validated double sandwich ELISA is used for the detection of anti-RVA antibodies in serum of guinea pigs immunized with RVA vaccines. Briefly, 96-well ELISA plates are sensitized with a hyperimmune anti-RVA serum obtained in cattle (colostrum-deprived calf, experimentally infected and hyperimmunized with bovine RVA). Plates are incubated for 18 hours at

4–8°C, and after a blocking step, clarified supernatant of MA-104 cultures infected with RVA with viral titer not less than 10^7 FFU/ml is added, and titer by ELISA detection of Ag of 1/100 (positive wells) or supernatant of MA-104 uninfected cells (negative wells) is required. Next, in both + and – coated wells, the serum samples are added in 6 4-fold serial dilutions. A commercial peroxidase-labeled anti-guinea pig IgG conjugate is then added. The reaction is developed using H_2O_2/ABTS as the substrate/chromogen system. The plates are read at 405 nm.

Coating buffer, washing buffer, and development buffers are the same as described for IBR ELISA.

Capture antibody: a hyperimmune anti-rotavirus group A serum made in colostrum-deprived calves is used with an anti-RVA Ab titer not less than 1/65,536.

Positive antigen: suspension of bovine rotavirus produced in MA104 cells. The virus to be used must be clarified and titrated in this system to determine its use dilution. Use viral stocks that have an infectious titer of 107 FFU/ml or greater and at least one positive signal by ELISA for Ag detection in the 1/100 dilution.

Negative antigen: Clarified supernatant of uninfected MA-104 cells.

Guinea pig anti-IgG detector antibody labeled with peroxidase

The assay has been validated using these two commercial reagents:

Affinity purified goat anti-Guinea Pig Ig G (H+L) peroxidase labeled, KPL, cat # 14-17-06

Peroxidase-conjugated affiniPure goat anti-guinea pig IgG(H+L), Jackson, cat# 106-035-003

Controls

In each plate, a positive control serum is used in a fixed dilution to be determined for each batch, a serum negative control that runs in a single dilution of 1/16, and a reaction blank (phosphate buffered saline [PBS]). Each of these samples is run in duplicate on each reaction plate.

Guinea pig positive control: Pool of sera from five guinea pigs vaccinated with two doses of vaccine containing 107 FFU/ml of RVA in oily adjuvant (reference vaccine). This serum must have an expected anti-RVA IgG Ab titer of 1/16,384, and in the dilution of use in the assay (1:1024), should give a corrected optical density (ODc) included in the following admission range:

Arithmetic mean ODc positive control $\pm$ 1 SD = 0.535 $\leq$ average of two replicates in the working dilution 1:1024 $\leq$ 0.871

Those plates in which the positive control falls within the range of acceptance are considered good. The average of the ODc of the positive control in the plate is considered as 100% positivity percentage (PP) for the calculation of the Ab titers of each sample.

Guinea pig negative control: pool of normal sera from adult guinea pigs not vaccinated with RVA whose corrected absorbance in the dilution of use is lower than the cut-off point of the technique (11% of the corrected optical density of the positive control). Run each assay plate in two replicates and a 1/16 dilution in each reaction plate.

Blank of reaction: PBS Tween20 0.05% washing buffer.

For each of the controls, four wells are used (two positive captures and their two negative captures).

ELISA procedure

1. *Plate coating*

Prepare the necessary volume of anti-RVA capture antibody in plate coating buffer, pH 9.6, according to the number of plates to use. Place 100 μl per well. Seal the plates with adhesive tape and incubate at 4°C for 18 hours.

2. *Blocking*

Discard the contents of the plate. Wash one time with wash buffer. Dry by hitting the plate against absorbent paper.

Add 100 μl per well of blocking buffer and incubate 1 h at 37°C in 5% CO_2 or in a humid chamber.

3. *Positive and negative Ag addition*

Discard the contents of the plate. Wash plate two times with wash buffer.

Make the working dilution of the positive antigen (Rotavirus Lot for ELISA) in washing buffer.

Dilute the negative Ag (MA-104 without infection), in the same dilution as the positive Ag, in washing buffer.

Place 100 μl of positive Ag in the odd columns and 100 μl of negative Ag in the even columns in each plate.

Incubate: 1 h at 37°C in 5% CO_2 or in a humid chamber.

Note: If the plates are not used at the time, at this step after adding the virus and after the washes, store at –20°C (±2°C) until use. Plates sensitized in this way can be used within 3°months of preparation.

4. *Dilution and plating of samples*

The serum samples are diluted in fourfold dilutions from 1.6 to 4.01 (six dilutions) in a surrogate plate. Each sample is tested from 1.6, 2.2, 2.8, 3.01, 3.4, 4.01. Transfer 100 μl of each dilution made to the reaction plate from the most dilute dilution to the more concentrated.

5. *Dilution and plating of controls*

The ELISA kit will be supplied with standardized controls that must be prepared at the working dilution: for example, positive control (1:1024) and negative control (1:16).

The controls (positive, negative, and PBS) are prepared in a tube according to the volume necessary for all the plates of the test, and they will be loaded directly on the original plate (100 μl per well).

1. Place 100 μl of the positive control dilution in wells A7, B7, A8, and B8.
2. Place 100 μl of the negative control dilution in wells A9, B9, A10, and B1.
3. Place 100 μl of PBS/0.05% Tween20 in wells A11, B11, A12, and B12 (reaction blank).

Incubate in a humid chamber for 1 hour at 37°C.

6. *Addition of peroxidase-conjugated detector antibodies*

1. Discard the contents of the plate. Wash four times. Dry off.
2. In a tube containing 10 ml of blocking buffer, place the corresponding amount of conjugated antibody according to its dilution of use.
3. Place 100 μl per well on the entire plate.
4. Incubate in a humid chamber for 1 hour at 37°C.
5. Discard the contents of the plate. Wash five times. Dry off.

7. Development, Reading, and Interpretation

1. Prepare 10 ml of developing solution. Place 100 μl of the development solution in each well and wait between 10 and 15 minutes with the plate in the dark. Take a reading of test at 405 nm to check that the positive control is reaching its expected optical density value.
2. Stop the reaction by placing 50 μl of stop solution (SDS 5%) in all wells of the plate and read. Transfer the reading data to a spreadsheet.
3. Perform the subtraction of each of the absorbances of the positive captures minus each of their respective negatives. Example: H1 minus G1 = ODc (Corrected Optical Densities).
4. Calculate the average of the ODc of the positive control (100% PP). Calculate the cut-off of each plate as 11% of the ODc value of the positive control. Calculate the average of the replicates of the negative control and its PP%.
5. Calculate the PP% of each sample in each dilution: PP = ODc sample / ODc Positive Control * 100.

Test acceptance (conformity)

The criteria described in the following will be applied individually to each plate.

An assay plate is considered compliant when the ODc of the positive control is within the established range. The negative control and the reagent blank show PP% lower than the cut-off of the assay (11% PP). The serum titer included as standard gives the expected value ± a fourfold dilution (method error). The antibody titer of a sample is established as the $\log_{10}$ of the inverse of the maximum dilution whose PP% is greater than or equal to the cut-off of the trial (11% PP).

Reporting results of immunogenic quality of rotavirus vaccines tested in guinea pigs and sera evaluated by ELISA

The results can be interpreted and used to classify a vaccine in the INTA guinea pig model if the ELISA run has been approved and there are a minimum of five animals with results to calculate the average titer of Ab induced by the vaccine.

To declare the immunization trial compliant, sera from guinea pigs immunized with the reference vaccine should yield an average titer within the established range determined by a control chart that shows the average value ± two standard deviations obtained from a minimum of five tests.

For the vaccine to be evaluated, the average anti-RVA Ab titer determined by ELISA is reported as the average of the $\log_{10}$ titers of at least five of the six immunized animals.

Negative samples in the lowest serum dilution tested (1/16) are expressed with an arbitrary titer of 0.3 for calculations.

According to the cut-off points of the guinea pig model for RVA, the vaccines can be classified as having low, intermediate, good, and very good immunogenicity (see Table 3.4).

Harmonization of assays for the region

A positive and negative control serum panel and reference vaccines will be elaborated and made available for regional users upon request to harmonize the results obtained for each assay laboratory adopting the control method.

4 Guinea Pig Model to Test the Potency of Bovine Parainfluenza Type 3 Virus Vaccines

Development and statistical validation of a guinea pig model and its associated serology assays as an alternative method for potency testing of bovine parainfluenza type-3 virus vaccines.

4.1 Introduction

Bovine parainfluenza type 3 virus (PI-3) is a member of the genus Respirovirus (Makoschey and Berge 2021) of the subfamily Paramyxovirinae, order Mononegavirales, family Paramyxoviridae. The viral genome is single-stranded non-segmented negative-sense RNA. Viral particles are spherical to pleomorphic, 150–200 nm in diameter, and consist of a nucleocapsid surrounded by a lipid envelope derived from the plasma membrane of the cell from which it buds. In this envelope, two viral glycoproteins are present: the hemagglutinin-neuraminidase (HN) and the fusion (F) glycoprotein, which mediates viral attachment to and penetration of the host cell. These glycoproteins represent the main viral antigens and induce protective antibody responses in infected animals. Hemagglutination, hemadsorption, hemolysis, and fusion are biological activities associated with these viral glycoproteins (Murphy et al. 1994).

PI-3 virus has been recognized as an endemic agent in the cattle population worldwide. The virus was first isolated in the United States from the nasal discharge of cattle with shipping fever (Ellis 2010). Clinical disease due to PI-3 infection is highly variable, from asymptomatic infection to severe respiratory disease and pneumonia characterized by cough, pyrexia, and nasal discharge. The clinical disease generally occurs in naïve calves with a low level of maternal passive antibodies or in animals under stress conditions. Lung lesions and immunosuppression after PI-3 infection contribute to the establishment of secondary bacterial infections (*Mannheimia haemolytica* and mycoplasma spp.), which are a common feature of enzootic pneumonia in calves and the bovine respiratory disease complex in feedlot

DOI: 10.1201/9781003373148-4

cattle, leading to severe bronchopneumonia (Ellis 2010; Haanes, Guimond, and Wardley 1997). The resulting disease is part of the bovine respiratory disease complex (BRDC) together with the bovine respiratory syncytial virus. The BRDC is considered one of the most significant illnesses associated with feedlot cattle in the United States and possibly worldwide (Snowder et al. 2006).

Regarding genetic and antigenic characterization, bovine PI-3 is classified into three genotypes: genotype A, mainly distributed in the United States and Europe, genotype B, circulating in Australia, and genotype C, only reported in China (Wen et al. 2012; Zhu et al. 2011; Makoschey and Berge 2021). In Argentina, the virus was detected in cases of respiratory disease in bovines and buffaloes. The strains found in bovines were classified as genotypes A and C, while the strains detected in buffaloes were typed as genotype B. Argentina is the first country so far to report the circulation of the three genotypes (Maidana et al. 2012).

Specific Abs induced in the infected animals possess the property to block viral hemagglutinin (HA) function. These Abs target specific HA antigens involved in the binding to red blood cells that can be measured by hemagglutination inhibition assay (HIA), a rapid and economical technique, which does not require complicated infrastructure and can be easily implemented in veterinary laboratories to evaluate the protective antibody responses to PI-3. This technique is a useful tool to conduct serologic surveys in the field and to evaluate vaccine potency in the target species as well as in a laboratory animal model. Animals exposed to PI-3 (after infection/vaccination) significantly increase their HI Ab titers. For viral agents within the Orthomyxoviridae and Paramyxoviridae families, the HI Ab titer in serum is associated with protection against infection (Ellis 2010; Benoit, Beran, and Devaster 2015).

There are numerous multivalent vaccines containing PI-3 on the market to prevent the BRDC. Vaccines are formulated with the live attenuated or inactivated virus. Vaccines containing inactivated PI-3 are formulated in aqueous or oil adjuvant together with other viral (bovine herpesvirus 1 [BoHV-1], bovine viral diarrhea virus [BVDV], and bovine respiratory syncytial virus [BRSV]) and bacterial antigens. It was postulated that a 1/32 titer of HI of passive maternal Abs in calves is the threshold of protective immunity against PI-3 infection (Ellis 2010). In the method we used to test vaccine potency in guinea pigs and cattle, this Ab titer expressed as hemagglutination inhibition units (HIU) is $32 \times 8 = 256$ HIU; $\log_{10}$ transformed = 2.4. To our knowledge, a unified criterion to evaluate the potency of killed PI-3 vaccines in the region has not yet been established.

In the present chapter, we report the statistical validation of a guinea pig model as a method for potency testing of inactivated PI-3 vaccines. As explained in the previous chapter, the validation involved the study of the kinetics of the Ab response in the laboratory animal model and the target species, a regression analysis applied to the dose–response curve to define categories for vaccines qualification, and a concordance analysis of vaccine testing in parallel in the model versus the natural host to demonstrate the predictive value of the Ab titer in guinea pigs to predict the vaccine potency in cattle.

4.2 Dose–response analysis: discrimination capacity of the guinea pig model and the target species (cattle)

Two sets of four polyvalent water-in-oil vaccines containing increasing concentrations of the PI-3 reference strain SF4 (Genotype A) covering a range from 10^5 to 10^8 tissue culture infectious units ($TCID_{50}$)/dose, together with fixed concentrations of BoHV-1 and BVDV, were formulated. Viral antigens were inactivated using binary ethylenimine (BEI) (Bahnemann 1990). The vaccines were formulated using pilot-scale equipment under industrial conditions. The adjuvant used in the formulation consisted of a mix of 0.67% polysorbate 80, 2.1% sorbitan 80 monooleate, and 57.9% mineral oil in a 60:40 oil:water proportion. The vaccines were applied following the same time intervals and dose volumes as commercial vaccines (30 days apart and 3 ml). These "reference vaccines" of known antigen concentration were used to estimate the dose–response to study the kinetics of Ab responses to PI-3 in bovine calves and guinea pigs. Each set of reference vaccines was tested in two independent field trials in bovine calves and two independent experiments in guinea pigs (Table 4.1).

It is important to note that bovine PI-3 infection is endemic in cattle worldwide and that all calves receive colostral anti-PI-3 antibodies that can persist until they are up to 6 months old. It is estimated that when colostral Ab titers decrease to values lower than 1/32 (256 HAI units), calves become highly susceptible to PI-3 infection, which may or may not be symptomatic, and rapidly develop antibodies against PI-3 (Ellis 2010). Considering this, the proposed dose–response vaccine sets 1 and 2 were tested in four independent field trials in Angus, Hereford, and their crossbreeds' calves, older than 6 months of age, belonging to three different beef farms. In all cases, the initial Ab titers against PI-3 were determined by HIA to assemble the groups with statistically similar mean Ab titers. The animals included in the trial were prime-vaccinated with the vaccines under study. Groups of 10 animals were vaccinated with 2 doses of 3 ml of each vaccine 30 days apart. Serum

Table 4.1 Dose–Response for PI-3 Vaccines. Experimental Design

Vaccine set 1	PI-3 SF4 concentration ($TCID_{50}$/dose)	Bovine field trial (1.1 and 1.2), *n*	Guinea pigs	Vaccine set 2	PI-3 SF4 concentration ($TCID_{50}$/dose)	Bovine field trial (2.1 and 2.2), *n*	Guinea pigs
A	10^8	10	10	C	10^7	10	10
B	5×10^7	10	10	C	10^7	10	10
C	10^7	10	10	D	10^6	10	10
D	10^6	10	10	E	10^5	10	10
Placebo/ unvaccinated control	Adjuvant media	5 + 5	5 + 5	Placebo/ unvaccinated control		5 + 5	5 + 5
Total		**50**	**50**			**50**	**50**

[a] Each vaccine set was tested in five animals (set 1 in five animals and set 2 in five animals).

[b] Each vaccine set (1 and 2) was tested in two independent assays in guinea pigs.

samples were taken at 0, 30, 60,and 90 post-vaccination days (pvd). A group of bovines receiving placebo (n = 5) and a group of non-vaccinated animals (n = 5) were included as controls. The entire study involved 100 bovines (Table 4.1). In parallel, guinea pigs weighing 400–500 grams were controlled to determine that they were seronegative for HIA Ab against PI-3 and then were vaccinated with two doses of each vaccine in a volume corresponding to one-fifth of the bovine dose (0.6 ml), 21 days apart. The animals were sampled at 0, 30, and 60 pvd. The entire study involved 100 guinea pigs.

Initially, the kinetics of the Ab dose–responses in both species was evaluated. Statistical analysis basically addressed the comparison of averages of the Ab titer through time from 0 to 60/90 pvd among the different doses. Data on HI Ab titers in guinea pigs and cattle were analyzed using a mixed model of repeated measures (Little and Raghunathan 1999). Differences between means were tested using the Bonferroni criterion. The Akaike criterion was used for the selection of the covariance matrix (Harada et al. 2010). In the cases in which the vaccine*time interaction was significant, the comparison of means was made within each time.

All bovine groups started the study with statistically similar initial pre-vaccination Ab titers against PI-3. The results showed that the vaccines formulated with 10^6 or higher $TCID_{50}$/dose were highly immunogenic, inducing significant increases in the Ab titers after one dose at 30 pvd, reaching a plateau that was maintained up to 90 pvd. At all timepoints, vaccinated groups differed statistically from the placebo and unvaccinated groups. In contrast, very low responses were induced in the animals receiving the vaccine formulated with 10^5 $TCID_{50}$/dose: after two doses of vaccine, the mean Ab titer did not differ from the placebo at 60 pvd and returned to the starting Ab titers by 90 pvd. The bovines were able to discriminate only between vaccines formulated with 10^5 and 10^6 $TCID_{50}$/dose or higher (Table 4.2a; Figure 4.1a).

Table 4.2a Dose–Response Study in Bovines: Mean HI Antibody Titer to PI-3

Vaccine	PI-3 concentration ($TCID_{50}$/dose)	*n*	Time (post-vaccination days)			
			0	30	60	90
A	1×10^8	10	2.41 B	3.32 A*	3.38 A*	3.33 A*
B	5×10^7	10	2.41 B	3.41 A*	3.41 A*	3.41 A*
C	1×10^7	30	2.14 B	3.27 A*	3.28 A*	3.22 A*
D	1×10^6	20	2.19 B	3.08 A*	3.13 A	2.89 A*
E	1×10^5	10	2.02 B	2.40 B	2.55 B	2.11 B
Placebo/ unvaccinated		20	2.10 B	2.17 B	2.12 B	2.01 B

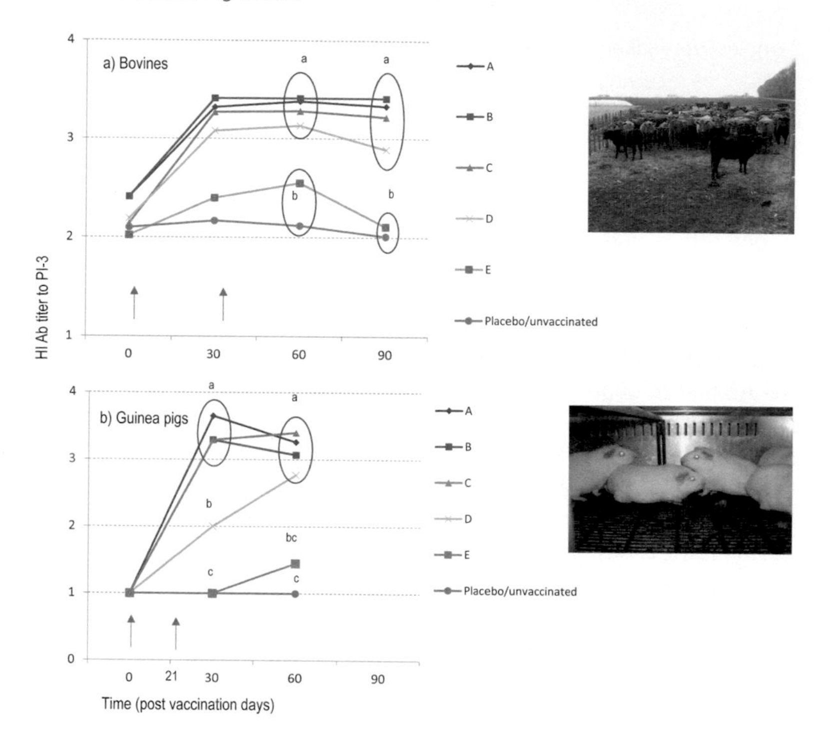

Figure 4.1 Kinetic antibody responses to PI-3 determined by hemagglutination inhibition assay (HIA) after immunization in (a) bovines and (b) guinea pigs. Lines represent mean HI antibody titers induced by the dose–response vaccines tested. Arrows indicate vaccination time in each species (0–30 in bovines and 0–21 in guinea pigs). Lines at 30/60 pvd with different letters indicate statistical differences among mean antibody titers induced by the vaccines tested (mixed model for repeated measures and Bonferroni method for multiple comparison; $p < 0.05$).

Table 4.2b details the statistical analysis of the total groups of guinea pigs immunized with each vaccine, grouped by PI-3 antigen dose. The guinea pigs started the experience seronegative for HI Ab against PI-3. At 30 pvd, after two doses of vaccines (0 and 21 days), all the guinea pigs of the vaccinated groups with vaccines containing 10^8, 5×10^7, 10^7, and 10^6 $TCID_{50}$/dose of PI-3 seroconverted. Vaccines containing an antigen concentration of 10^7 $TCID_{50}$/dose or higher developed significantly higher Ab titers at 30 pvd than vaccines formulated with 10^6 $TCID_{50}$/dose, which in turn differ from vaccines containing 10^5 $TCID_{50}$/dose, which in practice do not induce a detectable response by HIA. In this last vaccine, the true mean difference barely exceeds, at most, half a log. The vaccine with 10^5 $TCID_{50}$/dose, as observed in cattle, fails to generate biologically relevant seroconversion in the four groups of vaccinated guinea pigs (Table 4.2b, Figure 4.1b).

Table 4.2b Dose–Response Study in Guinea Pig: Mean HI Antibody Titer to PI-3

Vaccine	PI-3 concentration ($TCID_{50}$/dose)	n	Days post-vaccination		
			0	30	60
A	1×10^8	25	1.00 c	3.65 a	3.26 a
B	5×10^7	9	1.00 c	3.29 a	3.07 a
C	1×10^7	30	1.00 c	3.29 a	3.40 a
D	1×10^6	19	1.00 c	2.00 b	2.77 a
E	1×10^5	10	1.00 c	1.00 c	1.45 bc
Placebo		20	1.00 c	1.00 c	1.00 c

Note: Values are expressed as arithmetic means of $\log_{10}$ transformed Ab titers. Global means of all bovines vaccinated with vaccine sets 1 and 2. Two independent trials for each vaccine set.
Means in the same column with different letters differ significantly.
Means in the same row with an asterisk indicate a significant difference (seroconversion with respect to the basal Ab level).
The Vaccine*Time interaction was significant (two way ANOVA repeated measures, Bonferroni multiple comparisons; $p < 0.0001$).

From the results obtained in these experiments in guinea pigs, it was concluded that the administration of two doses of vaccine with an interval of 21 days between doses allows a kinetics and magnitude of response of Ab anti-PI-3 similar to those observed in bovines to be obtained. The guinea pig model was able to discriminate among vaccines differing by 1 log of antigen concentration in the range between 10^5, 10^6, and 10^7 or higher at 30 pvd. Likewise, the guinea pig, probably due to its seronegative status, has greater discriminating power than the bovine to evaluate the immunogenicity of vaccines for PI-3, with 30 pvd being the optimal time for taking the sample to carry out the evaluation. The guinea pig model, like cattle, did not seroconvert when receiving vaccines with 10^5 $TCID_{50}$/dose; thus, this antigen concentration represents the detection limit of the model.

4.3 Regression analysis: selection of quality split points for PI-3 vaccines

To determine the split points for vaccine classification in both species, regression analysis was used. Briefly, for the anti-PI-3 HI Ab titers induced by the reference vaccines tested in the dose–response trials in each species (guinea pigs and cattle), a polynomial linear model of second degree was used. The reported F, the coefficient of determination (R^2), and the *p*-value corresponding to the significance of the model were estimated, showing an appropriate goodness of fit to the data for both species. These

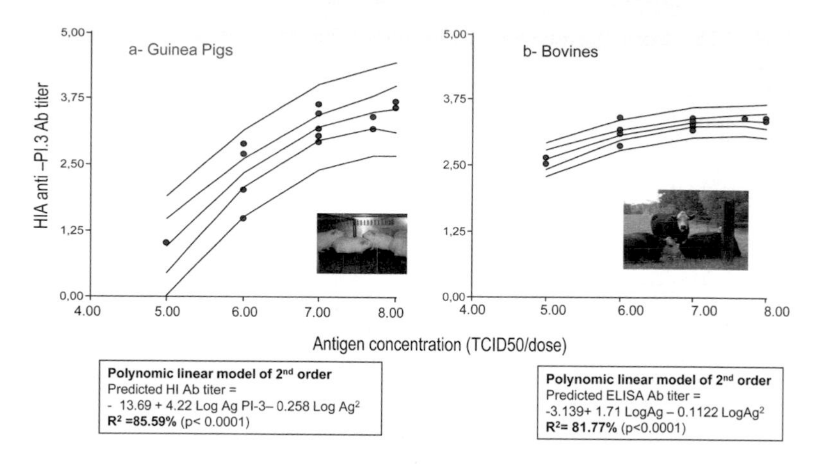

Figure 4.2 Regression analysis from which cut-off split points are estimated to classify vaccines of different immunogenicity (potency) in cattle and guinea pigs. Each point represents the average of the $\log_{10}$ of anti-PI-3 hemagglutination inhibitory Ab titers obtained in groups of five guinea pigs/cattle vaccinated with oil vaccines with increasing concentrations of PI-3 Ag. The curve is estimated by the model.

mathematical models allowed prediction of the mean anti-PI-3 HI Ab titers induced by the vaccines with 85% adjustment for guinea pigs and 81% adjustment for cattle. From the adjusted model, we selected two antigen concentrations (10^5 and 10^7 $TCID_{50}$/dose) to define three different categories of vaccine quality: very satisfactory, satisfactory, and non-satisfactory. The cut-offs were established and mean Ab titer given by the lower limits of the 90% prediction intervals (LIP) for the Ab titer expected for these concentrations. In each case, the estimated curves and their corresponding confidence and prediction bands are depicted in Figure 4.2.

The regression curve along with the bands corresponding to the 90% confidence and prediction intervals is depicted.

Table 4.3 details the prediction limits obtained from the mathematical model in cattle and guinea pigs. The lower limit of the 90% prediction interval represents the minimum mean HI Ab titer that a vaccine should induce for each concentration of Ag in the formulation when administered to a group of at least five guinea pigs or five cattle, with 95% confidence (coverage). Given that both the model and the target species were able to discriminate between vaccines formulated with a concentration of 10^6 Ag PI-3 or higher from vaccines with lower concentrations, the cut-off point to declare a vaccine as being of minimum acceptable quality was the corresponding 90% prediction limit for this antigenic concentration, estimated at a minimum titer of 2.8 for bovines and 1.5 for guinea pigs. Additionally, those vaccines

Table 4.3 Estimated 90% Prediction Limits from the Regression Analysis

		Bovine HI Ab titer		Guinea pig HIAb titer	Predicted Ab titers			
					IPRE(LI)	IPRE(LS)	IPRE(LI)	IPRE(LS)
Vaccine	PI-3 concentration (log_{10}DICT50/dose)	T0	T60	T30	Bov T60		CobT30	
A	8	2.69	3.41	3.59	3.05	3.67	2.67	4.46
B	7.7	2.32	3.41	3.41	3.09	3.67	2.67	4.34
C	7	2.38	3.29	3.18	3.05	3.62	2.39	4.02
C	7	2.26	3.35	3.18	3.05	3.62	2.39	4.02
D	6	2.44	3.11	1.48	2.80	3.37	1.52	3.17
A	8	2.14	3.35	3.71	3.05	3.67	2.67	4.46
B	7.7	2.51	3.41	3.17	3.09	3.67	2.67	4.34
C	7	2.63	3.23	2.93	3.05	3.62	2.39	4.02
D	6	2.57	2.87	2.02	2.80	3.37	1.52	3.17
C	7	2.08	3.17	3.47	3.05	3.62	2.39	4.02
C	7	1.66	3.29	3.05	3.05	3.62	2.39	4.02
C	7	2.26	3.41	3.65	3.05	3.62	2.39	4.02
C	7	1.72	3.35	3.47	3.05	3.62	2.39	4.02
D	6	2.14	3.41	2.88	2.80	3.37	1.52	3.17
D	6	1.72	3.17	2.69	2.80	3.37	1.52	3.17
E	5	2.2	2.63	1	2.28	2.93	0.02	1.89
E	5	1.84	2.51	1	2.28	2.93	0.02	1.89

Table 4.4 PI-3 Vaccine Classification Criteria by HI Assay in Guinea Pigs and Bovines

	Vaccine potency HI PI-3		
Species	**Non-satisfactory**	**Satisfactory**	**Very satisfactory**
Guinea pig	$\bar{y} < 1.5$	$1.5 \leq \bar{y} < 2.4$	$2.4 \leq \bar{y}$
Bovine	$\bar{Y} < 2.8$	$2.8 \leq \bar{Y} < 3.1$	$3.1 \leq \bar{Y}$

Cut-offs represent the Ab titer to PI-3 determined by HIA, expressed as the $\log_{10}$ of the hemagglutination units (HIU) obtained in the serum of the vaccinated animals. Arithmetic mean Ab titer of groups of five guinea pigs, evaluated 30 pvd, and groups of five bovines, evaluated 60 pvd. Bovines receive two doses of vaccine with a 30-day interval, following the vaccine manufacturer's recommendations, and are sampled at 0 and 60 pvd. This latter point corresponded to the peak or plateau of Ab titers reached by aqueous or oil vaccines, respectively.
Guinea pigs receive two doses of vaccine (one-fifth the volume of the bovine dose) with a 21-day interval and are sampled at 0 and 30 pvd. The two-dose regimen chosen in the laboratory animal model allows the immune response induced by vaccines of low potency to be detected. The 21-day interval between doses was adopted in order to obtain a curve of Ab kinetic response similar to that obtained in bovines, but in a shorter period of time, providing a faster alternative method for vaccine potency testing than the one conducted in bovines.

with antigen concentrations of 10^7 $TCID_{50}$/dose or higher should yield mean Ab titers of no lower than 3.1 in bovines and 2.4 in guinea pigs (Table 4.4).

A classification criterion to evaluate the immunogenicity of PI-3 commercial vaccines in bovines and guinea pigs was obtained. Vaccine with titers higher than 2.4 in guinea pigs (3.1 in cattle) are highly immunogenic and considered to be of very satisfactory potency, while vaccines not reaching an average titer of 1.5 are considered to be of low immunogenicity. An intermediate value would indicate that the vaccine has low PI-3 antigen titers but that it could be immunogenic in already primed animals, so it would be considered to be of satisfactory potency.

4.4 Concordance analysis between bovines and guinea pigs for vaccine classification

To study the degree of concordance between the guinea pig model and cattle, a total of 71 cases were analyzed (Table 4.5).

Using the established cut-off points, the level of agreement in vaccine classification between the guinea pig model and cattle was evaluated using the weighted kappa index, accompanied by its asymptotic standard error

Table 4.5 Vaccines Included in the Concordance Analysis

Type of vaccine	Vaccine composition (antigens)	PI-3 concentration ($TCID_{50}$/dose)	Syndrome	Adjuvant type	Number of vaccines tested[1]	Number of comparative assays[c]
Calibration vaccines for the dose–response curve[a]	PI-3-IBR-BVDV	10^8, 10^7,10^6, 10^5	Respiratory	Oil	11	17
Gold standard vaccine of known concentration[b] 6	IBR gE-1,	No PI-3	Reproductive	Oil	1	6
	PI-3-IBR-BVDV	10^7–10^8	Respiratory	Aqueous	4	2
	PI-3, IBR, BVDV	10^7	Respiratory	Aqueous + saponin	1	2
Commercial vaccines[c] 14	PI-3-IBR-BVDV	Unknown	Multipurpose	Oil	4	4
	PI-3-IBR-BVDV	Unknown	Pre-service	Aqueous	2	2
	IBR	No PI-3	Queratoconjuntivitis	Aqueous	1	1
	PI-3-IBR- BVDV, *n* = 5, + BRSV, *n* = 6	Unknown	Respiratory/pneumonia	Oil	14	14
		Unknown	Respiratory/pneumonia	Aqueous	11	11
Placebo/ unvaccinated controls					16	16
Total						**75**

[a] Vaccines detailed in Table 3.1 and Figure 3.1.
[b] Gold standard vaccines: vaccines of known antigen concentration (potency) prepared under industrial conditions.
[c] Commercial vaccines of unknown antigen concentration representing the available vaccines in the local market.

(SE) and the 95% confidence interval (Vanbelle 2016; Viera and Garrett 2005). In a first analysis, this index was estimated considering a single cut-off point that differentiates satisfactory from unsatisfactory vaccines (1.5 guinea pigs; 2.8 cattle) and using 39 comparison lines corresponding to gold standard vaccines and negative controls (17 vaccines formulated for the dose–response assay, 6 gold standard vaccines of known Ag concentration, and 16 negative controls). The degree of concordance of the guinea pig model and the target species was almost perfect (weighted kappa = 0.898). Only two vaccines were classified as unsatisfactory by the cattle and satisfactory by the guinea pig model. Both vaccines correspond to aqueous formulations with 10^7 $TCID^{50}$/dose that were appropriately classified by the guinea pigs (Table 4.6).

In a second analysis, 32 commercial vaccines of unknown PI-3 concentration were included in the analysis. An optimal or substantial degree of concordance (weighted kappa = 0.777) was obtained. Only 8 discrepancies out of a total of 71 comparisons were registered (Table 4.7).

Finally, the third analysis was carried out considering two cut-off points to discriminate between satisfactory and very satisfactory vaccines. The degree of concordance was also very good (weighted kappa= 0.757, Table 4.8), and it was concluded that the guinea pig model is well validated to predict the Ab titer that PI-3 vaccines will induce in the target species.

4.5 Discussion and conclusions

The dose–response study conducted in guinea pigs and cattle with oil-adjuvanted killed vaccine for PI-3 vaccines showed similar kinetics and

Table 4.6 Concordance between the Guinea Pig Model and Bovines (Gold Standard Vaccines)

	Bovines (2.8)		
Guinea pigs (1.5)	**Not satisfactory**	**Satisfactory**	
Not satisfactory	**19**	0	19 (48.7%)
Satisfactory	2	**18**	20 (51.3%)
	21 (–53.80%)	18 (–46.0%)	39
Weighted kappa[a]	**0.898**		
Standard error	0.07		
95% confidence interval	0.760 to 1.000		

Diagonal values are in agreement between the models, which are indicated in bold.

Table 4.7 Concordance between the Guinea Pig Model and Bovines (Gold Standard Vaccines and Commercial Vaccines)

	Bovine (2.8)		
Guinea pig (1.5)	**Not satisfactory**	**Satisfactory**	
Not satisfactory	**33**	0	33 (46.5%)
Satisfactory	8	**30**	38 (53.5%)
	41 (57.7%)	30 (42.3%)	71
Weighted kappa[a]	**0.777**		
Standard error	0.072		
95% confidence interval	0.635 to 0.919		

Diagonal values are in agreement between the models, which are indicated in bold.

Table 4.8 Concordance between the Guinea Pig Model and Bovines Considering Two Split Points and Three Categories of Immunogenicity

	Bovine			
Guinea pig	**Not satisfactory**	**Satisfactory**	**Very satisfactory**	
Not satisfactory	**34**	0	0	34 (47.9%)
Satisfactory	4	**7**	2	13 (18.3%)
Very satisfactory	3	4	**17**	24 (33.8%)
	41 (57.7%)	11 (15.5%)	19 (26.8%)	71
Weighted kappa[a]	**0.757**			
Standard error	0.064			
95% confidence interval	0.632 to 0.882			

Diagonal values are in agreement between the models, which are indicated in bold.

magnitude of the HI Ab responses for both animal species. A directly proportional relationship between Ab response and antigen concentration was observed in both species. However, the guinea pig model showed a clearer dose–response effect than in cattle and was able to discriminate the potency of vaccines differing by 1 log_{10} in antigen concentration.

A classification criterion to evaluate the immunogenicity of PI-3 commercial vaccines in bovines and guinea pigs was obtained, showing a substantial agreement to properly classify the potency of commercial vaccines.

The guinea pig model can be used to test the batch-to-batch quality of PI-3 vaccines to be released on the market and represents a practical tool for vaccine companies as well as the animal health authorities to guarantee optimal products on the market.

The guinea pig model to control the vaccine potency of PI-3 vaccines was chosen as an American Committee for Veterinary Medicines (CAMEVET) guideline in 2015 and was used to survey the quality of the vaccines in Argentina (see Chapter 6). Recently, in order to have a more convenient tool to test the vaccines, an enzyme-linked immunosorbent assay (ELISA) was developed to measure the Ab responses in cattle and guinea pigs (Maidana et al. 2021). Validation of the guinea pig model using this new ELISA technique in comparison with the results obtained by HIA will be conducted in order to transfer the model to the animal health services with the gold standard technique and the ELISA, a more versatile technique that can be available in a commercial format.

Chapter 4 Annex III

Potency test for PI-3 vaccines in the guinea pig model

Guidelines elaborated by PROSAIA biologic ad hoc group and approved as CAMEVET harmonized document in 2015: www.prosaia.org/biologicos-veterianrios/, https://rr-americas.woah.org/es/proyectos/camevet/documentos-armonizados/?se=parainfluenza&page-nb=1

PI-3 Hemagglutination inhibition test

This assay determines the presence of antibodies (Abs) directed to the bovine PI-3 viral hemagglutinin in the serum of infected and/or vaccinated animals. Prior to the assay, serum is treated with kaolin to adsorb unspecific inhibitors of the hemagglutination. The sample is also treated with red blood cells (RBCs) to absorb unspecific hemagglutinating substances present in the serum. Treated serum will end in a 1/5 dilution. Serial twofold dilutions (5, 10, 20, 40, etc.) of the serum are mixed with a fixed concentration of PI-3 virus established as 8 HA units/25 µl. The reaction is developed by adding guinea pig RBCs. In positive sera for Abs directed to the viral hemagglutinin, the formation of the Ag–Ab complex will inhibit PI-3 RBC hemagglutination, and RBCs will aggregate, forming a red button in the bottom of the well. The end point of the hemagglutination inhibition activity of a serum sample determines its HI Ab titer to PI-3 and is expressed as the reciprocal of the highest dilution of serum in which complete hemagglutination did not occur. This value multiplied by the constant 8 (representing virus concentration) defines the PI-3 hemagglutination inhibition units (HIU) of the sample.

Materials

- U-bottom 96-well plates
- Cuvettes
- 200-µl and 1000-µl tips
- 1.5-, 15-, and 50-ml plastic tubes
- Pasteur pipettes
- Needle ″25/8
- 5-ml syringe

Equipment

- Mono-channel micropipettes: up to 50 µl, 200 µl, 1000 µl
- 8–12 Multi-channel micropipettes 5–50 µl
- Microcentrifuge (up to 14,000 rpm)
- Refrigerated centrifuge (up to 5000 rpm)

Reagents

- Virus: PI-3 viral suspension produced in MDBK cells containing 8 HAU/25 μl (16 HAU/50 μl)
- Phosphate buffer: pH 7.2–7.4 (PBS)
- Kaolin solution:
  Kaolin 0.04 g
  PBS 5 ml
- Guinea pig RBC suspension
- Alsever anticoagulant
- **Positive control serum:** pool of sera from guinea pigs immunized with 2 doses (21 days apart) of a reference vaccine containing a 10^7 $TCID_{50}$/dose of PI-3 in oil adjuvant, sampled at 30 pvd. Selection of serum with 320–640 HIU is recommended.
- **Negative control serum:** pool of sera from non-immunized guinea pigs.

Guinea pig RBC suspension

1. Take a simple of blood with anticoagulant (1 ml anticoagulant + 4 ml of blood) from a guinea pig by cardiac puncture. Blood extraction must be conducted under anesthesia following European Centre for the Validation of Alternative Methods (ECVAM) recommendations for animal welfare.
2. Discard the needle and pour the blood into a 15-ml tube.
3. Centrifuge the blood at 1500 ± 200 rpm for 5 ± 1 min at 4–8°C.
4. Discard supernatant and wash twice with PBS in the same manner.
5. Prepare a 0.8% RBC working suspension:
   a. Suspension 1/4 = 1 ml of RBC + 3 ml PBS
   b. Suspension 1/120 = 1 ml of ¼ suspension + 29 ml PBS
6. Count RBCs in the Neubauer chamber and adjust the final suspension to contain 5 × 10^7 cells/ml.

IMPORTANT: the RBC suspension should be prepared fresh at the moment of running the HI assay. After each wash, the supernatant should remain clear. The presence of a reddish color is an indication of hemolysis, and the RBCs are not suitable for the assay.

PI-3 virus titration by hemagglutination assay (HA)

1. Thaw the PI-3 virus.
2. Add 50 μl PBS to all wells of a U-bottom 96-well plate.
3. Add 50 μl of virus to four wells (four replicates): 1-A, B, C, D.
4. Perform serial twofold dilution, transferring 50 μl from 1 to 12.
5. Add to the plate 50 μl of the 0.8% RBC suspension.

6. Incubate at room temperature (20–27°C) for 1 hour.
7. Once the red buttons are present in the control wells, the assay is ready to read.
8. The virus for the assay should have 16 HA/50 μl (8 HA/25 μl). If the obtained titer is lower than this, the viral suspension is not suitable for the assay. If the viral suspension has a higher titer, perform a suitable dilution in PBS and repeat the titration.

Treatment of serum samples prior to HI testing

1. Heat serum at 56°C for 30 min.
2. In a 1.5-ml test tube, mix 50 μl of serum with 50 μl of Kaolin, vortex, and incubate for 10 ± 2 min at room temperature (20–27°C).
3. Centrifuge for 15 ± 2 min at 1500 rpm.
4. Take 50 μl of the supernatant and transfer to a new tube with 75 μl of PBS. (Final serum dilution: 1/2 × 2/5 = 1/5.)
5. Add 10 μl of RBC suspension, incubate with gentle agitation at 37°C for 30 min, and centrifuge for 15 ± 2 min at 1500 rpm. Transfer the supernatant to a new tube. Run the HI assay thereafter or save the treated sample at –20°C until used. Use within 1 week.
6. Positive and negative control sera are treated in the same manner.

HI assay

1. Add 25 μl PBS to all wells of a U-bottom 96-well-plate except row G.
2. Add 25 μl of the treated serum to wells 1-F, G, H (sample 1), 2-F, G, H (sample 2), and so on.
   - Vertical design: 12 samples are tested from row 1 to 12 in 7 serial dilutions (G–A).
   - Horizontal design: 8 samples are tested from row A to H in 11 dilutions (2–12).
3. Perform serial twofold dilutions, transferring 25 μl from F to A.
4. Run positive and negative standard sera of known HI titer in the assay, randomly mixed among the samples and plates.
5. In the last plate of the assay, include a positive and negative control serum and PBS as reaction blank (RBC control). Confirm the titration of the virus.
6. Add to the plates 25 μl of the working dilution of the PI-3 virus (8 HA/25 μl) in all the rows except H. This well will serve as control of serum sample + RBC. The absence of a red button in this well indicates that the serum sample still possesses unspecific hemagglutination activity, and its HI titer will not be determined.
7. Incubate the plates with the serum–virus mix for 1 hour at room temperature (24–27°C).

8. Add 50 μl of the 0.8% RBC suspension to all the plates. Incubate at room temperature. Once the red buttons are present in the control wells, the assay is ready to read by direct inspection.

Assay approval

An HI assay is approved if:

- The back titration of the working virus suspension gives an HA titer of 16 HA/50 μl (8 HA/25 μl)
- The RBC button in the control wells (RBC + PBS) is solid
- The standard and control sera give the expected HI titers (± 1 twofold dilution).

Assay interpretation

The HI titer is considered the reciprocal of the highest dilution of serum in which complete agglutination did not occur. The titer is expressed in hemagglutination units (HIU), multiplying the reciprocal of the final dilution by the number of HA units of the virus (8 HA). See Table 4.9.

Vaccine potency testing in guinea pigs

Validation criterion for guinea pig testing

Potency testing in guinea pigs is considered valid when the mean Ab titer obtained from animals vaccinated with a reference vaccine of satisfactory potency reaches the expected value (higher than 1.50 in immunized guinea pigs and higher than 2.80 in bovines) and unvaccinated control animals remain seronegative for Ab against PI-3 throughout the experience.

Calculation

All sera of animals immunized with the vaccine under control will be evaluated. FIVE (5) sera taken at 30 pvd with the highest Ab titers to PI-3

Table 4.9 HI Ab Titer against PI-3

Well (vertical design)	Reciprocal of the serum dilution in the well	Serum HI units (reciprocal dil*8)	Ab HI titer to PI-3 ($\log_{10}$ HIU)
G	5	40	1.6
F	10	80	1.9
E	20	160	2.2
D	40	320	2.5
C	80	640	2.8
B	160	1280	3.1
A	320	2560	3.4

(expressed as the log_{10} transformed of HIU) will be selected, and an average will be calculated on that basis.

Vaccine approval criterion by potency testing in guinea pigs
For the approval of the vaccine submitted to control, mean Ab titers obtained must be higher than or equal to 1.50.

Harmonization of assays for the region

A panel of positive and negative control sera and reference vaccines should be elaborated and made available for regional users to harmonize the results obtained for each assay laboratory adopting the control method. The reference vaccine will allow the conformity of each immunization assay to be defined, while the panel of reference sera will be used to validate the results of the serologic assays (HI test) and for the standardization of alternative assays (ELISA, viral neutralization [VN]).

5

Guinea Pig Model to Test the Potency of BVDV Vaccines

Development and statistical validation of a guinea pig model and its associated serology assays as an alternative method for potency testing of bovine viral diarrhea virus vaccines.

5.1 Introduction

Bovine viral diarrhea virus (BVDV) belongs to the genus Pestivirus, within the family Flaviviridae. BVDV is classified into two genotypes: BVDV-1 and BVDV-2. BVDV-1 is divided into 17 subgenotypes, while BVDV-2 is divided into 3 subgenotypes (Ridpath 2010, 2012). Recently, a new pestiviral species was proposed, known as "HoBi-like-virus" or "BVDV-3" (Bauermann and Ridpath 2015). BVDV strains are differentiated into genotypes through phylogenetic analyses and serotypes using monoclonal antibodies (Vilcek et al. 1997). Moreover, BVDV can present two biotypes: non-cytopathogenic (NCP) and cytopathogenic (CP), classified according to the visible effects (lesions) that it generates in epithelial cell cultures. Generally, the NCP biotype is the one that circulates most frequently in livestock populations (Ridpath 2010, 2012).

In Argentina, the data reported by Odeón and colleagues, based on a serological survey carried out in three regions of the country (Buenos Aires, La Rioja, and Corrientes), revealed a high prevalence (48.6%–90.7%) of antibodies against BVDV-1a in adult cattle (older than 2 years of age), indicating the wide distribution and exposure to this virus in bovine cattle from these states of the country (Odeón et al. 2001). In agreement with these data, more recent reports describe 79.8% (77.9%–80.8%) prevalence in adult animals from breeding herds of the Chubut province (Pecora and Perez Aguirreburuald 2017).

Regarding the viral variants circulating in Argentina, BVDV-1b is the most frequent subgenotype (reaching 70% of isolates), followed by subgenotype 1a and genotype 2 (Pecora et al. 2014; Spetter et al. 2021). These percentages vary by country; in Brazil, 80% of the isolates analyzed in the last published epidemiological studies belonged to BVDV-1a, BVDV-2b,

DOI: 10.1201/9781003373148-5

and HoBi-like virus (Flores et al. 2018). Recently, a phylogenetic study was reported in Mexico, in which a predominance of BVDV-1c subgenotype was found among 60 BVDV isolates (Gomez-Romero et al. 2017, 2021a). In the last phylogenetic study carried out in Uruguay, most of the BVDV isolates were grouped in subgenotype 1a, although BVDV-1i and VDVB-2b isolates were also found (Maya et al. 2020). In Chile, similar percentages of VDVB-1 and VDVB-2 isolates have been found (Aguirre, Quezada, and Celedon 2014; Alocilla and Monti 2022; Donoso et al. 2018). In the United States, some studies indicate that BVDV-1b is the most frequent subgenotype in bovine herds (Fulton 2015; Ridpath et al. 2011). Although BVDV is a ubiquitous virus, several countries from the European Union have implemented plans for its control/eradication based on the detection and elimination of persistently infected (PI) animals (Schweizer et al. 2021; OIE 2018).

5.1.1 Acute infection

This is the classic form of the disease. The most likely route of transmission is through the oropharyngeal and respiratory tracts. The animal can be infected with a CP or NCP biotype and develop a respiratory, digestive, or reproductive presentation. It may result in subclinical infection or severe disease with high mortality that may be accompanied by thrombocytopenia and hemorrhagic diathesis, depending on the genotype of the virus (Ridpath et al. 2011, 2006). Respiratory disease may be one of the manifestations of generalized BVDV infection, and some investigators have associated it with strains of subgenotype 1b (Fulton 2015). The tropism of this virus by cells of the immune system, infecting white cells and causing depletion of circulating B and T lymphocytes, results in temporary immunosuppression that favors subsequent infections with other pathogens (Bolin 1995). There is epidemiological and experimental evidence that BVDV is directly associated with the bovine respiratory complex (BCR), interacting with pathogens such as bovine respiratory syncytial virus (BRSV); infectious bovine rhinotracheitis-1 (IBR); *Mannheimia haemolytica*; *Mycoplasma bovis*; parainfluenza-3 (PI-3), etc. (Grissett, White, and Larson 2015).

5.1.2 Congenital infection

The consequences of BVDV fetal infection depend on the time of exposure to the virus during gestation and are associated with transplacental dissemination during viremia. If the fetal infection occurs from 0 to 45 days of gestation, it can cause embryonic death, while an infection from day 45 to day 175 of gestation can cause miscarriage, birth defects, and/or immunotolerance. If the infection occurs between days 125 and 285, it can cause abortion and birth of weak calves, although normal calves are also born with pre-colostral neutralizing antibodies against BVDV (Malacari et al. 2018). Venereal infection can occur when using acutely infected or PI bulls. The infection of the semen with BVDV occurs because the virus replicates in the seminal vesicle and in the prostate gland. Infected semen presents a decrease in sperm motility and an increase in the percentage of morphological abnormalities of spermatozoa. An occasional risk factor is the presence of non-viremic bulls with persistent testicular infection, which continuously excrete viruses in the semen (González Altamiranda et al. 2012; Ridpath et al. 2006a, b).

5.1.3 Persistent infection

PI animals are the product of a transplacental infection with NCP-BVDV during gestation between 35 and 125 days. It is believed that the requirement for the generation of persistence is exposure to the virus when the fetal immune system is in development (40–125 days). PI animals are immunotolerant to BVDV homologous viruses, but they may be immunocompetent to heterologous strains of BVDV and also to other infectious agents such as BoHV, PI-3, etc. (Lanyon et al. 2014). BVDV persists in all the tissues of the PI animals, especially in cells of the immune system. Persistent BVDV infection is lifelong, and PI calves are considered the main disseminators of the infection in the herds (Donoso et al. 2018; Flores et al. 2018; González Altamiranda et al. 2012; Ridpath et al. 2006a).

5.1.4 Mucosal disease

Mucosal disease (MD) is a sporadic manifestation of BVDV infection, which occurs only in PI animals when they are over infected with the CP biotype. The presence of the CP biotype may be external, by superinfection with homologous CP biotypes (Ridpath 2010; Ridpath et al. 2006a), or due to molecular rearrangements (mutation) in the persistent NCP strain (e.g. deletions, insertions, duplications of viral sequences, and insertions of host sequences), thus causing the CP strain. The disease is characterized by severe clinical signs (including severe leucopenia, profuse diarrhea, erosions and ulcerations of the whole digestive system), low morbidity, and a lethality rate close to 100%. MD may present an acute or chronic course, depending on the homology between the BVDV NCP and the CP affecting the PI animal (Bolin 1995).

5.1.5 Hemorrhagic syndrome

This syndrome, associated with acute infections with BVDV-2 strains, generates bloody diarrhea, epistaxis, conjunctival and mucosal congestion, petechial and ecchymotic mucosal hemorrhages, pyrexia, leucopenia, lymphopenia and neutropenia, and very high mortality rates (Ridpath et al. 2006a).

5.1.6 Serological tests

The antibodies present in the bovine sera as a product of infection or vaccination can be measured by classical viral neutralization (VN) tests or by enzyme-linked immunosorbent assay (ELISA). Positive and negative controls must be included in each test. The determinations are considered valid if the controls are within the expected limits. To carry out the VN assay, CP-BVDV strains are used in order to simplify the reading of the results under an optical microscope. Some widely used CP strains are "Singer", "Oregon C24V", and "NADL", which belong to genotype 1 of the BVDV. Given the antigenic variability detected among BVDV strains, when prevalence studies are performed, special attention should be given to the BVDV used: it should present sufficient antigenic similarity with the viral variants that circulate in the local cattle (Pecora et al. 2014). On the other hand, in the case of experimental vaccine efficacy/immunogenicity studies, VN assays should be performed using the same strain as was used in the vaccine formulation or a strain that has sufficient homology to the vaccine strain (OIE 2018).

Regarding the ELISA tests to detect antibodies to BVDV available on the market, most of them are based on the detection of antibodies against conserved viral proteins (such as NS3 or Erns). This strategy overcomes the restrictions previously described for the VN assay in terms of antigenic variability among strains; however, it is still a limited tool to confirm the free status of an animal (Gonda et al. 2012; Kampa et al. 2007).

5.1.7 Vaccines

Vaccines against BVDV can be of two types: those formulated with attenuated virus and those based on inactivated virus. Attenuated virus vaccines generally give a more consistent immune response as compared with killed vaccines. However, the use of these vaccines may present several risks: in the first place, they cannot be used in pregnant females, since they can produce fetal infection and abortion. Likewise, vaccination of PI animals with attenuated strains could induce MD. Ultimately, as with any live vaccine, there may be risks of a poorly attenuated strain or reversion to the virulence of the vaccine strain with undesired effects (Balint et al. 2005).

Inactivated virus vaccines are considered much safer than attenuated vaccines, but they need to be administered in several doses, and also, the use of adjuvants or immunomodulators is required to obtain satisfactory levels of immunity (Griebel 2015).

Experimental subunit vaccines have been developed for BVDV, based on the E2 glycoprotein produced in various expression systems, such as baculovirus in insect cells, or mammalian cells. A targeted subunit vaccine was recently launched on the market in Argentina with excellent results (Bellido et al. 2021; Pecora et al. 2012, 2016). This subunit vaccine is as safe as the killed vaccines and induces very high Ab responses, comparable to those of live attenuated vaccines, without the risk of potential infections. In addition, the subunit vaccine also acts as a DIVA (Differentiating Infected from Vaccinated Animals) vaccine.

To our knowledge, there is no unified criterion for batch-to-batch potency testing of inactivated BVDV-containing vaccines in their formulation in the region. There are specific recommendations for the immunogenicity and efficacy of the killed BVDV vaccines in the American Code of Federal Regulation (9CFR 113.215 2014), which indicates that any vaccine containing inactivated antigen of BVDV (killed virus or subunit vaccine) to be registered in the United States needs to be tested in seronegative calves. Five vaccinated and three control seronegative calves must be included in the test. Fourteen days or more after the last vaccination, according to the manufacturer's recommendation, blood samples are taken and serum VN Ab responses tested. The negative controls must remain seronegative. If at least four of the five vaccinated calves in a valid test have not developed 50% endpoint VN Ab titers of 1:8 or greater, the serial is unsatisfactory and must go under challenge. The vaccinates and controls may be challenged with virulent BVDV. The animals shall be observed for 14 days post challenge. If two or more vaccinates show a temperature of 104.0°F for 2 or more days and develop respiratory or clinical or other signs, the vaccine is unsatisfactory. The World Organisation for Animal Health in the Manual of Diagnostic Tests and Vaccines for Terrestrial Animals recommends similar tests (OIE 2018).

Since the BVDV infection is endemic in cattle worldwide, access to seronegative bovines is quite difficult; therefore, the development of a statistically well-validated laboratory animal model represents a very useful tool to evaluate the potency of combined and monovalent vaccines to control BVDV.

In the present chapter, we report the statistical validation of a guinea pig model as a method for potency testing of inactivated and subunit BVDV vaccines. As explained in previous chapters, the validation involves the study of the kinetics of the Ab response in the animal model and the target species, a regression analysis applied to the dose–response curve to define categories for vaccine qualification, a concordance analysis between the laboratory animal model and the natural host to classify the vaccines, and a BVDV calf challenge model to correlate the Ab titers with protection against infection and disease.

5.2 Dose–response assay analysis: discrimination capacity of the guinea pig model and the target species (cattle)

The standardization of the guinea pig model for BVDV started in 2003. In 2008, after analyzing the results of several pilot trials (2003–2005), a dose–response trial was proposed with a set of four oil and four aqueous trivalent vaccines (IBR, BVDV, and PI-3), as detailed in Table 5.1. The BVDV strain used in the formulation was Singer 1a. The oil-based vaccines were administered in doses of 3 ml in cattle and 0.6 ml in guinea pigs, while the aqueous vaccines were administered in doses of 5 ml in cattle and 1 ml in guinea pigs.

The vaccines were evaluated in guinea pigs in two independent assays involving 120 animals. The animals received two doses of 0.6 ml or 1 ml of oil or aqueous vaccines, respectively, 21 days apart. The vaccines were evaluated in beef calves older than 6 months of age, seronegative for VN antibodies to BVDV 1a, and preferably from herds free of BVDV infection. Cattle were vaccinated with two doses of vaccine at an interval of 30 days. Serum samples were taken at the start of the experiment, at the time of each dose, and then up to 60 post-vaccination days (pvd). The complete study of the dose–response trial of oil-adjuvanted vaccines included a total of 100 bovines. The aqueous vaccine trial involved 50 bovines.

The mean anti-BVDV Ab VN titer obtained for each vaccine tested in each species, corresponding to groups made up of 4 to 5 guinea pigs and between 4 and 10 cattle, vaccinated with the calibrator vaccines formulated with increasing concentrations of BVDV in oily and aqueous formulations were used:

i) To study the kinetics and magnitude of the neutralizing antibody response of oil and aqueous formulations in each species to decide the best timepoint to compare the Ab titer in guinea pigs versus cattle.

Table 5.1 BVDV Reference Vaccines Prepared for the Dose–Response Study

Oil-adjuvanted vaccines		Aqueous-adjuvanted vaccines	
Formulation	BVDV concentration ($TCID_{50}$/dose)	Formulation	BVDV concentration ($TCID_{50}$/dose)
O-1	10^8	**A-1**	5×10^8
O-2	10^7	**A-2**	5×10^7
O-3	10^6	**A-3**	5×10^6
O-4	10^5	**A-4**	5×10^5

ii) To study the discriminating ability of the guinea pig model and bovines to differentiate between vaccines with different BVDV antigenic concentrations in the context of both oil and aqueous adjuvants.
iii) To set up quality split points to evaluate the potency of the vaccines in both animal species.

From the study of data of 17 oil-adjuvanted vaccines for cattle and 24 for guinea pigs, the cut-off points were established that allow the qualities of these vaccine formulations to be determined. The regression curve of aqueous vaccines included 7 vaccines tested in cattle and 11 vaccines tested in guinea pigs.

The humoral immune response against vaccination with BVDV virus antigen was evaluated in the serum of vaccinated animals using VN, fixed virus (100 $TCID_{50}$)–variable serum. Sera were tested from an initial 1/4 dilution and in serial two-base dilutions to follow World Organisation for Animal Health (OIE) guidelines. The VN Ab titer is expressed as the log_{10} of the inverse of the maximum dilution of serum with neutralizing activity (Annex IV).

5.2.1 Dose–response of oil BVDV vaccines in guinea pigs and bovines

At first, we evaluated the kinetics of the VN Ab responses to oil vaccine in guinea pigs up to 60 pvd. Given the lack of normality and variance homogeneity (caused by the lack of response of the group immunized with 10^6 $TCID_{50}$ per dose; variance = 0), the statistical analysis was performed using a general mixed linear model, where the trial was considered as a fixed effect (block) and pigs as a random variable within each block. The model also assumes heterogeneity of variance, which can be modelated. Groups of guinea pigs vaccinated with 10^8 $TCID_{50}$ per dose developed Ab titers significantly higher than those vaccinated with 10^7 $TCID_{50}$ per dose at 30 and 60 pvd, demonstrating an optimal discriminating capacity between the three concentrations evaluated. The obtained results indicated that in this working range, the guinea pig model was capable of significantly discriminating between vaccines formulated with doses differing by 1 log_{10}. The earliest timepoint for vaccine evaluation can be at 30 pvd. In addition, at least 1 $\times$ 10^7 $TCID_{50}$ of BVDV antigen per dose is needed to induce a detectable neutralizing Ab response, since the four groups of five guinea pigs each that received the vaccine formulated with 1 $\times$ 10^6 $TCID_{50}$ per dose did not develop detectable Ab against BVDV by the methodology used (Figure 5.1).

The evaluation of oil-adjuvanted vaccines in cattle that were included in the dose–response analysis comprised four sets of vaccines prepared in the years

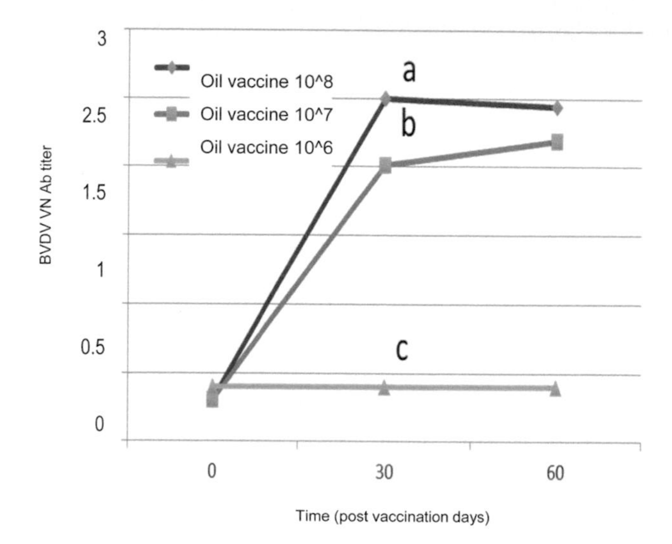

Figure 5.1 Dose–response of groups of guinea pigs immunized with the oil vaccines formulated with increasing doses of Ag. Time 30 and 60 dpv points with different letters, A, B, or C, differ significantly (general linear mixed model, $p < 0.05$).

2003, 2004, 2008, and 2015 with antigen concentrations covering a range from 1×10^5 to 1×10^8 $TCID_{50}$/dose. The vaccines were applied to beef calves of the Aberdeen Angus and Hereford breeds and their crosses, older than 6 months of age, *previously selected as seronegative for Ab against BVDV.* In turn, the animals included in the trials had to be primed with the vaccines under study. The bovine experiments included between four and eight animals per vaccine. The animals were vaccinated with two doses of 3 ml of vaccine at an interval of 30 days, and serum samples were taken at 0, 30, and 60 pvd. Groups of non-vaccinated animals were included in all cases as controls. The entire study achieved utilizable results from 96 vaccinated cattle.

In cattle immunized with the oil-adjuvanted vaccines, at both 30 and 60 days, after one and two doses of vaccine, seroconversion was noted in the groups vaccinated with the vaccines formulated with 1×10^7 and 1×10^8 of BVDV Ag per dose, while the vaccines formulated with 1×10^6 and 1×10^5 did not induce detectable levels of neutralizing Ab against BVDV in seronegative cattle. These doses were not enough to prime naïve animals, in agreement with the results obtained in the lab animal model (Table 5.2, Figure 5.2).

The mean Ab titers of cattle immunized with the 1×10^8 Ag vaccines were significantly higher than the titers induced by the 1×10^7 vaccines (repeated

Table 5.2 Dose–Response Assay of Oil-Adjuvanted BVDV Vaccines in Cattle

Vaccine BVDV concentration ($TCID_{50}$/dose)	Time (post-vaccination days)				
	0 (first dose)	30(second dose)	60	*n* Bovines	Herds
O-1 10^8	0.27 D	2.17 AB	2.54 A	8	1
O-2 10^7	0.51 D	1.41 C	2.15 B	31	4
O-3 10^6	0.50 D	0.50 D	0.50 D	44	5
O-4 10^5	0.32 D	0.31 D	0.29 D	8	1

O-1 to O-4 mean oil vaccines. Mean in the same column with different uppercase letters differs significantly (repeated measure ANOVE through time).

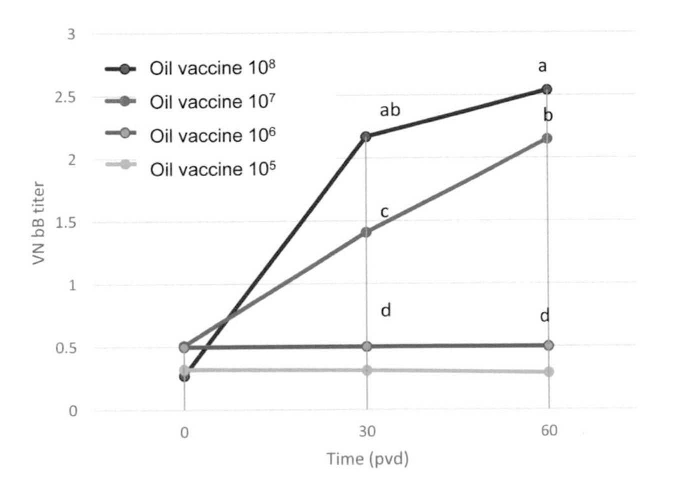

Figure 5.2 Dose–response of groups of calves immunized with the aqueous vaccines formulated with increasing doses of Ag. Time 30 and 60 dpv points with different letters differ significantly (general linear mixed model, two-way ANOVA of repeated measures through time, $p < 0.05$).

measures analysis of variance [ANOVA] through time, $p < 0.05$), indicating that the target species, like guinea pigs, is capable of discriminating between oil-adjuvanted vaccines formulated with virus concentrations of 1 log_{10} difference in the range under study.

5.2.2 Dose–response of aqueous BVDV vaccines in guinea pigs and bovines

The aqueous vaccines included in the dose–response study were also evaluated in two independent experiments in guinea pigs. The study of the kinetics of the VN Ab responses indicated that the guinea pig model was able to differentiate between aqueous vaccines formulated with concentrations

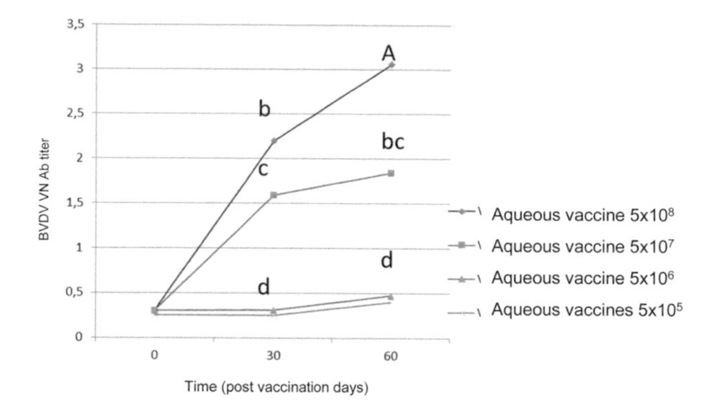

Figure 5.3 VN Ab Response in guinea pigs immunized with aqueous BVDV vaccine. Time 30 and 60 dpv points with a different letter differ significantly (general linear mixed model, two-way ANOVA of repeated measures through time, $p < 0.05$).

that differ by 1 $\log_{10}$ at both 30 and 60 pvd. Again, for aqueous formulations, it is noted that a virus concentration in the order of 10^7 infectious doses or higher was necessary to achieve detectable neutralizing antibody titers (Figure 5.3).

The aqueous vaccines evaluated in cattle that were included in the dose–response studies corresponded to the set prepared in 2008 (Table 5.1). These vaccines were formulated with five times more virus than the oil-adjuvanted vaccines (1×10^8 oil versus 5×10^8 aqueous) and were tested in seronegative cattle belonging to two beef herds. In turn, the animals included in the trials had to be primed with the vaccines under study. The bovine experiments included between four and eight animals per vaccine. The animals were vaccinated with two doses of 5 ml of vaccine at an interval of 30 days, and serum samples were taken at 0, 30, and 60 pvd. Groups of non-vaccinated animals were included in all cases as negative controls. The entire study achieved valid results from 30 vaccinated cattle.

Cattle immunized with the formulations containing 5×10^8 and 5×10^7 $TICD_{50}$/dose seroconverted after the first and second doses and achieved slightly higher antibody titers than their oil counterparts formulated with five times less virus (Table 5.3; Figure 5.4).

Interestingly, the group of cattle that received the aqueous version with 5×10^6 $TICD_{50}$/dose of BVDV developed a neutralizing Ab response, while its oil counterpart with 1×10^6 did not induce detectable levels of Ab by the methodology used. This result was striking, since guinea pigs immunized with the same formulation did not develop a detectable VN Ab response. The results

Table 5.3 Kinetics of VN AB Responses to BVDV Vaccines in Cattle

Vaccine	BVDV concentration ($TCID_{50}$/dose)	VN Ab titers pvd 0	21	30	60	*n*
Aqueous-adjuvanted	5×10^8	0.30	1.96	2.78	2.82	8
	5×10^7	0.30	0.87	1.96	2.56	8
	5×10^6	0.30	0.60	0.83	1.96	4
Oil-adjuvanted	1×10^8	0.30	1.25	2.18	2.56	8
	1×10^7	0.30	0.69	1.09	1.69	8
	1×10^6	0.30	0.30	0.30	0.30	4
Placebo		0.30	0.30	0.30	0.30	8

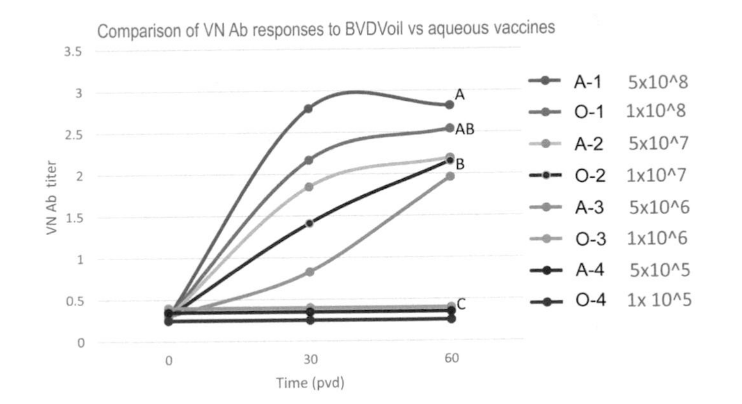

Figure 5.4 Comparison of the response of neutralizing Ab against BVDV in cattle immunized with aqueous and oily vaccines.

obtained suggest that both types of formulations can generate an optimal immune response and that it is dependent on the antigenic concentration and the immune status of the bovines (infection-free animals, or seronegative for 1a but having been in contact with a different variant of BVDV).

At 60 pvd, after two doses of vaccine, statistical analysis using the Kruskal–Wallis non-parametric rank sum test of the Ab titers reached by each group of vaccinated cattle indicated that the vaccine containing 5×10^8 in aqueous adjuvant was the one that induced the highest neutralizing titers, followed by the group receiving the 1×10^8 oil vaccine with intermediate Ab titers. The vaccines containing 5×10^7, 1×10^7, and 5×10^6 formed a group with a similar Ab titer among them but significantly lower than the group composed of the aqueous 5×10^5 and the oil 1×10^6 and 1×10^5 and vaccines that did not induce a detectable VN Ab response (Kruskal–Wallis non-parametric test, $p < 0.05$; Table 5.3 and Figure 5.4).

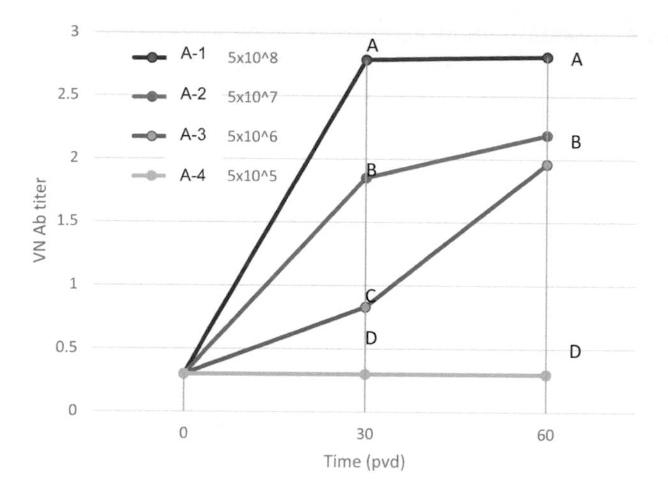

Figure 5.5 Response of neutralizing Ab against BVDV in cattle immunized with aqueous vaccines.

When the aqueous vaccines were analyzed among themselves but separately from the oil vaccines, it was observed that the bovines were able to discriminate among them at 30 pvd (after the first dose) and at 60 pvd (after two doses). In this way, it was decided to maintain the 60 pvd timepoint to compare with the guinea pig model at 30 pvd, in agreement with the timepoint selected to test the potency of these combined vaccines for IBR and PI-3 (Chapters 2 and 4) (Figure 5.5).

5.3 Regression analysis: selection of quality split points for BVDV vaccines

The data obtained from previous pilot studies and the dose–response trials carried out in cattle and guinea pigs were used in several regression analyses, which allows cut-off or split points to classify the vaccines according to their potency (immunogenicity) in guinea pigs and cattle to be established.

5.3.1 Linear regression analysis of oil vaccines in guinea pigs and bovines

To establish quality split points, as in the previous chapters, a total of 24 mean VN Ab titers from guinea pigs and 17 from groups of bovines vaccinated with the oil-adjuvanted calibrator vaccines were analyzed by linear regression analysis. The analysis for the lab animal model included experiments carried out between 2003 and 2016, involving a total of 80 guinea pigs (Table 5.4). The regression analysis for the target species included trials

Table 5.4 Mean Anti-BVDV VN Ab Titers of Guinea Pigs Vaccinated with Calibrator Oil Vaccines Formulated with Increasing Doses of BVDV Ag That Were Included in the Linear Regression Analysis. Predicted Mean VN Ab Titers and IP90% Given by the Mathematical Model

Assay number	BVDV concentration (Log_{10} $TCID_{50}$/ dose)	VN Ab titer (30 pvd)	Mean VN Ab titer predicted value	90% prediction lower level	90% prediction upper level
1	8	2.83	2.71	**2.01**	2.29
2	8	2.18	2.71	**2.01**	2.29
3	8	2.33	2.71	**2.01**	2.29
4	8	2.41	2.71	**2.01**	2.29
5	8	2.59	2.71	**2.01**	2.29
6	7	1.65	1.61	**1.37**	1.69
7	7	1.56	1.61	**1.37**	1.69
8	7	2.29	1.61	**1.37**	1.69
9	7	1.69	1.61	**1.37**	1.69
10	7	1.99	1.61	**1.37**	1.69
11	7	2.03	1.61	**1.37**	1.69
12	7	2.18	1.61	**1.37**	1.69
13	7	2.17	1.61	**1.37**	1.69
14	7	1.44	1.61	**1.37**	1.69
15	7	1.48	1.61	**1.37**	1.69
16	6.7	1.26	1.29	1.01	1.33
17	6.7	1.23	1.29	1.01	1.33
18	6.7	1.43	1.29	1.01	1.33
19	6	0.30	0.52	−0.15	0.13
20	6	0.30	0.52	−0.15	0.13
21	6	0.30	0.52	−0.15	0.13
22	6	0.30	0.52	−0.15	0.13
23	6	0.30	0.52	−0.15	0.13
24	6	0.30	0.52	−0.15	0.13

The values in bold correspond to the lower prediction limit of the cutoff point for very satisfactory vaccines (2.01); 1.37 is the cutoff point for satisfactory vaccines.

conducted from 2004 to 2016, involving 85 bovines seronegative for BVDV (Table 5.5).

For both species, the mathematical model that best fitted the data was a linear second-degree polynomial model (Figure 5.6). The mathematical model was able to predict the average VN anti-BVDV Ab titers induced by oil vaccines with a 92% adjustment for guinea pigs and a 94% adjustment for cattle (Figure 5.6).

Table 5.5 Mean Anti-BVDV VN Ab Titers of Cattle Vaccinated with Calibrator Oil Vaccines Formulated with Increasing Doses of BVDV Ag That Were Included in the Linear Regression Analysis. Predicted Mean VN Ab Titers and IP90% Given by the Mathematical Model

Assay number	Field trial	BVDV concentration (log_{10} $TCID_{50}$/dose)	*n* Bovines	VN Ab titer at 60 pvd	Mean VN Ab titer predicted	90% Prediction lower level	90% Prediction upper level
1	7	8	6	3.00	3.29	**2.13**	2.42
2	1	8	4	2.41	3.29	**2.13**	2.42
3	2	8	4	2.71	3.29	**2.13**	2.42
4	8	7	5	2.16	2.61	**1.54**	1.89
5	8	7	7	2.19	2.61	**1.54**	1.89
6	1	7	4	2.18	2.61	**1.54**	1.89
7	2	7	4	1.65	2.61	**1.54**	1.89
8	3	7	4	2.57	2.61	**1.54**	1.89
9	4	7	4	2.37	2.61	**1.54**	1.89
10	1	7	5	1.42	2.61	**1.54**	1.89
11	1	6.7	5	0.30	0.84	−0.24	0.11
12	2	6.7	5	0.30	0.84	−0.24	0.11
13	1	6	5	0.30	0.84	−0.24	0.11
14	1	6	5	0.30	0.84	−0.24	0.11
15	1	6	5	0.30	0.84	−0.24	0.11
16	1	6	5	0.30	0.84	−0.24	0.11
17	1	6	5	0.30	0.84	−0.24	0.11

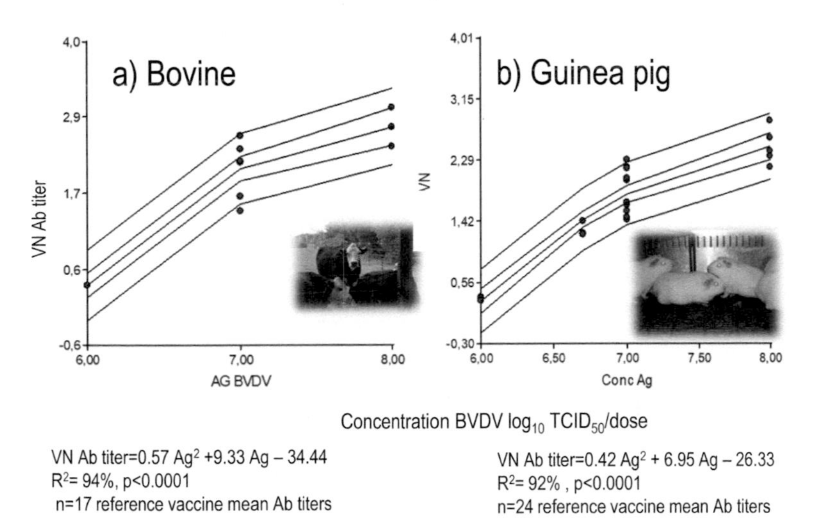

Figure 5.6 Result of the regression analyses in cattle (a) and guinea pigs (b) vaccinated with oil-adjuvanted reference vaccines. Observed means (blue dots), curve of means predicted by the model with the 90% corresponding confidence and prediction intervals.

The mean VN Ab titers predicted for each vaccine dose, together with the 90% prediction lower and upper limits obtained from the mathematical model in guinea pigs and cattle, are detailed in Tables 5.4 and 5.5. The lower limit of the 90% prediction interval represents the minimum mean VN Ab titer that a vaccine must induce for each concentration of Ag in the formulation when it is administered to a group of at least five guinea pigs/cattle. Given that the target species only developed a detectable and sustained response to VN with vaccines formulated with a BVDV concentration of 10^7 $TCID_{50}$/dose or higher, the cut-off point for declaring a vaccine of minimum acceptable quality was the corresponding 90% prediction lower limit for that antigenic concentration estimated at the minimum mean neutralizing Ab titer of 1.37 in guinea pigs and 1.54 for cattle, while very satisfactory vaccines should exceed a limit of 2.01 in guinea pigs and 2.13 in cattle.

5.3.2 Linear regression analysis of aqueous vaccines in guinea pigs and cattle

For these analyses, results were available for 7 vaccines tested in cattle and 11 vaccines tested in guinea pigs (Tables 5.6 and 5.7).

The mathematical model fitted for aqueous vaccines was also a linear second-degree polynomial model (Figure 5.7). The mathematical model was able to predict the average VN anti-BVDV Ab titers induced by aqueous vaccines with a 96% adjustment for guinea pigs and 83% adjustment for

Table 5.6 Mean Anti-BVDV VN Ac Titers of Guinea Pigs Vaccinated with Calibrator Aqueous Vaccines Formulated with Increasing Doses of BVDV Ag That Were Included in the Linear Regression Analysis. Predicted Mean VN Ab Titers and IP90% Given by the Mathematical Model

Assay number	BVDV concentration (log_{10} $TCID_{50}$/dose)	VN Ab titer (30 pvd)	90% prediction	
			Lower level	Upper level
1	8.70	2.03	1.09	3.36
2	8.70	2.11	1.09	3.36
3	7.70	2.51	0.69	2.76
4	7.00	1.68	0.22	2.28
5	7.00	1.45	0.22	2.28
6	6.70	0.30	0	2.02
7	6.70	0.30	0	2.02
8	6.70	0.30	0	2.02
9	6.00	0.30	−0.63	1.38
10	5.70	0.30	−0.63	1.38
11	5.70	0.30	−0.63	1.38

Table 5.7 Aqueous Vaccines Tested in Cattle

Assay number	BVDV concentration (log_{10} $TCID_{50}$/dose)	VN Ab titer at 60 pvd	90% prediction lower level	90% prediction upper level
1	8.70	2.78	**1.48**	3.75
2	8.70	2.56	**1.48**	3.75
3	7.70	2.56	**1.31**	3.41
4	7.70	2.56	**1.31**	3.41
5	7.70	1.63	**1.31**	3.41
6	6.70	1.96	0.55	2.70
7	5.70	0.30	−0.88	1.71

cattle (Figure 5.7). The split points for guinea pig were 0.22 and 1.02, too low to be applicable from the immunological point of view, whereas, the split points for cattle were 1.31 and 1.48.

5.3.3 Linear regression analysis including oil and aqueous reference BVDV vaccines in guinea pigs and bovines

In order to find a consensus cut-off for vaccine classification, a final regression analysis was conducted including all the calibrator vaccines, independently of the adjuvant used for their formulation. The goodness of fit was 94% and 91%, respectively (Figure 5.8). The split points for guinea pigs

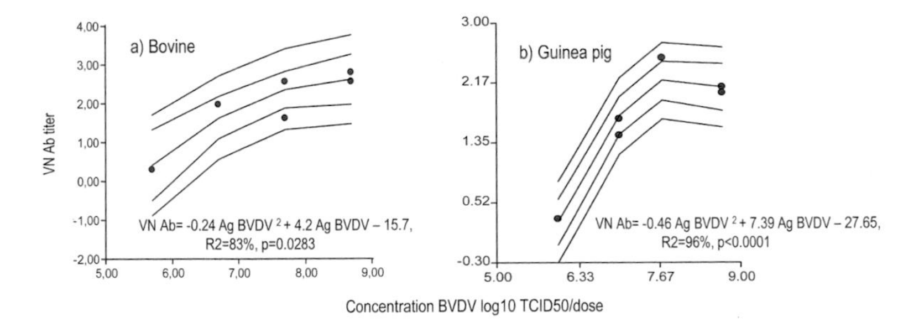

Figure 5.7 Result of the regression analyses in cattle (a) and guinea pigs (b) vaccinated with aqueous-adjuvanted reference vaccines. Observed means (blue dots), curve of means predicted by the model with the 90% corresponding confidence and prediction intervals.

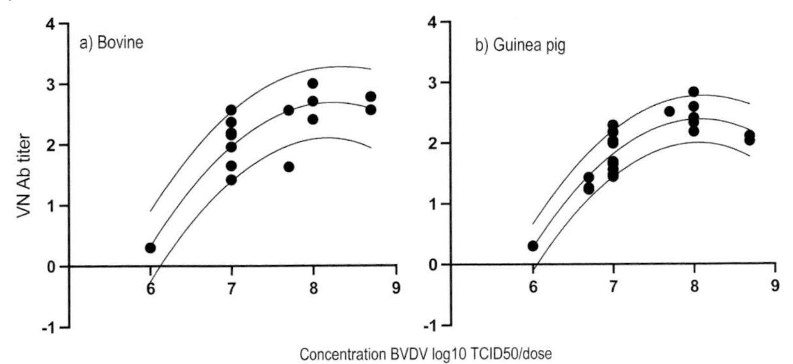

Figure 5.8 Result of the regression analyses in bovines (a) and guinea pigs (b) vaccinated with BVDV reference vaccines (oil and aqueous). Equation for bovines: VN = $-0.47\ Ag^2 + 7.78\ Ag - 29.35$, $R^2 = 91.18\%$; $p < 0.05$). Equation for guinea pigs: VN Ab = $-0.48\ Ag^2 + 7.86\ Ag - 29.36$, $R^2 = 94\%$; $p < 0.05$).

considering the lower limit of the 90% prediction interval were 1.43 and 1.99, while for cattle, they were 1.39 and 2.06.

5.3.4 Split points selected from the parametric statistical analysis on oil vaccines

The estimated sets of split points were evaluated in the concordance analysis (Table 5.8).

5.4 Concordance analysis between bovines and guinea pigs for BVDV vaccine classification

To study the degree of concordance between guinea pigs and bovines, a total of 86 comparison lines were analyzed, including 30 trials with calibrator

Table 5.8 Tentative Classification Points for BVDV Vaccines

Species	Statistical model	Type of reference vaccine included	*n*	BVDV vaccine potency		
				Unsatisfactory: less than 10^7 $TCID_{50}$/dose	**Satisfactory: 10^7–10^8 $TCID_{50}$/dose**	**Very satisfactory: 10^8 or higher $TCID_{50}$/dose**
Guinea pig	Linear regression second degree	Oil	24	$\bar{y} < 1.37$	$1.37 \leq \bar{y} \leq 2.0$	$2.0 < \bar{y}$
		Oil + aqueous	35	$\bar{y} < 1.43$	$1.43 \leq \bar{y} \leq 2.0$	$2.0< \bar{y}$
Bovine	Linear regression second degree	Oil	17	$\bar{Y} < 1.54$	$1.54 \leq \bar{Y} \leq 2.13$	$2.13 < \bar{Y}$
		Oil + aqueous	24	$\bar{Y} < 1.39$	$1.39 \leq \bar{y} \leq 2.08$	$2.08 < \bar{Y}$

Cut-off points expressed as the $\log_{10}$ titer of the highest reciprocal dilution of the serum of guinea pigs and cattle that presented neutralizing antibodies in vaccinated animals with the unknown vaccine. (y) Average titer of groups of five guinea pigs evaluated at 30 days post-vaccination (dpv); (Y) groups of five cattle evaluated at 60 dpv. The cattle received two doses of vaccine at an interval of 30 days and are sampled at 0 and 60 dpv. Guinea pigs received two doses of vaccine (one-fifth of the volume of the bovine dose) at an interval of 21 days and are sampled at 30 dpv.

vaccines (18 oil and 12 aqueous) prepared to carry out the dose–response study, 9 negative control groups receiving placebo or not vaccinated (negative controls), and 28 vaccines of known potency (gold standard), which included killed virus and two subunit vaccines based on the E2 protein and E2 fused to APCH single chain antibody (Bellido et al. 2021; Pecora et al. 2012). The commercial version of the latter form is named VEDEVAX; it was developed in the National Agricultural Technology Institute (INTA) and approved in the Argentinean market in 2018. Finally, 22 respiratory or reproductive commercial vaccines were included (Table 5.9). The trials included in the concordance analysis involved around 350 bovines and more that 300 guinea pigs.

In a first stage, the concordance study was carried out with a single cut-off point to differentiate satisfactory from unsatisfactory vaccines with the two sets of split points obtained in the prior section. The concordance analysis was initially carried out including the 86 comparison lines using the lower split points calculated in the previous section to discriminate between non-satisfactory and satisfactory vaccines.

As can be observed in Table 5.10, both split points gave very high kappa values, yielding in both cases almost perfect agreement between the guinea pig model and the target species to classify the potency of the BVDV vaccines. However, the split point obtained from the regression including the oil and aqueous reference vaccines yielded a slightly higher kappa index than the one calculated only with the oil reference vaccines. Thus, all further

Table 5.9 Vaccines Included in the Concordance Study. The Kappa Coefficient (κ), Used in This Analysis, Corresponds to the Proportion of Concordances Observed over the Total of Observations, Having Excluded the Concordances Attributable to Chance

Vaccine	Type	Aqueous	Oil	Comparison lines
Calibrator vaccines	Trivalent dose–response	4	5	12 + 18 = 30
Gold STD	Killed virus	1	12	28
	E2 subunit		2	
	APCH E2Exp.		7	
	VEDEVAX		7	
Commercial	Respiratory	14	3	22
	Reproductive	1	4	
Placebos/unvaccinated				9
Total				**86**

Table 5.10 Concordance to Classify BVDV Vaccines: Comparison between the Two Split Points

Comparison		**Bovine 1.54**	
		Not satisfactory	**Satisfactory**
Guinea pig	Not satisfactory	**33**	1
1.37	Satisfactory	6	**46**
Total			86
Kappa		**0.834**	
p-value		2.47×10^{-44}	

Comparison		**Bovine 1.39**	
		Not satisfactory	**Satisfactory**
Guinea pig	Not satisfactory	**32**	1
1.43	Satisfactory	5	**48**
TOTAL			**86**
Kappa		**0.8558**	
p-value		8.30×10^{-52}	

Diagonal values are in agreement between the models, which are indicated in bold.

Table 5.11 Oil Vaccines

Comparison		Bovine 1.39	
		Not satisfactory	**Satisfactory**
Guinea pig 1.43	Not satisfactory	**16**	1
	Satisfactory	3	**38**
Total			**58**
Kappa		**0.8391**	
p-value		1.75×10^{-27}	

Table 5.12 Aqueous Vaccines

Comparison		Bovine 1.39	
		Not satisfactory	**Satisfactory**
Guinea pig 1.43	Not satisfactory	23	0
	Satisfactory	2	10
Total			**35**
Kappa		**0.8679**	
p-value		4.66×10^{-22}	

comparisons were conducted with this set of split points, 1.39 for bovines and 1.43 for guinea pigs.

When the oil vaccines and the aqueous vaccine were analyzed separately, the split point yielded almost perfect agreement between the guinea pig model and bovines, indicating that the lab animal model was suitable to predict the potency of the BVDV vaccine in cattle independently of the adjuvant used (Tables 5.11 and 5.12). When the same analysis is done with the split points 1.37 for guinea pigs and 1.54 for bovines, the agreement was similar for oil vaccines (K = 0.844) but lower for aqueous vaccines(K = 0.798).

When the gold standard vaccines of known BVDV concentration, excluding the calibrator vaccines, were analyzed separately, almost perfect agreement was obtained with both sets of split points (Table 5.13). When the commercial vaccines alone were analyzed, substantial agreement is obtained with the split points 1.39/1.43 (Table 5.14, K = 0.7307), which was slightly better than that obtained with the cut-off 1.37/1.54 (K = 0.703). Finally, if the classification was carried out considering two split points and three vaccine categories (0: not satisfactory, 1: satisfactory, and 2: very satisfactory), there was substantial agreement between the guinea pig model and bovines (Table 5.15). The same analysis using the split points 1.37–2.00 for guinea pigs and 1.54–2.13 for bovines gave a slightly higher weighted kappa index (Table 5.16). We left it to each animal health authority to decide which cut-off they would like to use to start the control in their country.

Table 5.13 Gold Standard Vaccines

Comparison		Bovine 1.39/154	
		Not satisfactory	**Satisfactory**
Guinea pig 1.43/1.37	Not satisfactory	**8**	0
	Satisfactory	2	**24**
Total			**34**
Weighted kappa		**0.8496**	
p-value		8.31×10^{-17}	

Table 5.14 Commercial Vaccines

Comparison		Bovine 1.39	
		Not satisfactory	**Satisfactory**
Guinea pig 1.43	Not satisfactory	**20**	1
	Satisfactory	2	**6**
Total			**29**
Weighted kappa		**0.7307**	
p-value		1.75×10^{-17}	

Table 5.15 All Vaccines Considering Three Categories

Comparison		Bovine 1.39–2.08		
		0	**1**	**2**
Guinea pig 1.43–2.0	0	**32**	3	2
	1	4	**3**	9
	2	1	3	**32**
Total				**86**
Weighted kappa		**0.7141**		
p-value		4.26×10^{-36}		

0: not satisfactory; 1: satisfactory; and 2: very satisfactory.

Table 5.16 All Vaccines Considering Three Categories

Comparison		Bovine 1.54–2.13		
		0	**1**	**2**
Guinea pig 1.37–2.0	0	**32**	1	0
	1	5	**2**	10
	2	1	2	**33**
Total				**86**
Weighted kappa		**0.7437**		
p-value		5.86×10^{-44}		

0: not satisfactory; 1: satisfactory; and 2: very satisfactory.

5.5 Potency versus efficacy: relationship of antibody titers in guinea pigs and bovines with protection against experimental infection of seronegative calves

In order to have a first approximation of the association between the potency of the vaccines obtained in the guinea pig model and the protection in cattle, we will make reference to the standardization of a colostrum-deprived calf model of BVDV infection and disease using two viral strains (Malacari et al. 2016, 2018) and a small pilot study regarding the efficacy of a trivalent E2 subunit vaccine (Pecora et al. 2015). Twelve newborn male Holstein calves were removed from their mothers immediately after birth to avoid colostrum intake. Calves (mean weight at birth 32.13 ± 2.03 kg) were housed in biosecurity boxes (BSL2) and fed with a milk replacer free of immunological components and balanced food. All animals were negative for BVDV infection as determined by negative viral isolation and reverse transcription polymerase chain reaction (RT-PCR) in peripheral lymphocytes. They were also free of antibodies against BVDV at birth as assessed by serum neutralization test using BVDV type-1 (NADL strain) and BVDV type-2 VS253 strains.

Four calves were vaccinated with two doses (days 20 and 40 of life) of an experimental trivalent subunit vaccine (containing E2-APCH 1a, 1b, and 2) in oil adjuvant (Pecora et al. 2015). A group of eight non-vaccinated calves was used as negative control. On day 60 of life (20 days post re-vaccination), calves were divided into two groups and challenged with 5×10^6 $TCID_{50}$ of BVDV-1b 98/204 (group 1) or BVDV-2b 98/124 (group 2) (Table 5.17).

At challenge, the immunized calves developed neutralizing Ab to BVDV, with titers in concordance with the vaccine classification given by the guinea pig model, while the negative controls remained seronegative (Table 5.17).

After challenge, virus shedding in nasal swabs and leucocyte-associated-viremia were evaluated, and a clinical score was recorded daily for 14 days, considering depression, nasal and ocular discharges, diarrhea, anorexia, and rectal temperature, because it was already documented that one of the strains (BVDV-1b) generated a respiratory disease and the other (BVDV-2) gastrointestinal disease.

In the challenge group with BVDV 1b, the negative controls had a mean of 6 days of viremia and 4 days of nasal virus shedding and clinical scores between 7 and 10. The two vaccinated animals had no virus shedding or viremia, and the clinical score was drastically reduced, since one calf was fully protected and the other showed a clinical score of 3 because of

Table 5.17 VN Ab Titers in Guinea Pigs and Calves

Vaccine	n	Guinea pig VN Ab titer BVDV 1a at 30 pvd	Calves (n)	VN Ab titer at 60 pvd		Challenge (5 × 10^6 $TCID_{50}$)
				BVDV 1a	BVDV 1b	
Trivalent subunit vaccine	5	2.64	2	2.5	0.90	BVDV-1b 98/204
			2			BVDV-2b 98/124
Unvaccinated	5	0.30	4	0.30	0.30	BVDV-1b 98/204
			4			BVDV-2b 98/124

a muco-serous nasal secretion at 7 days post infection (dpi) and a slight depression at 8 dpi.

In the calves challenged with BVDV-2b (98/124), the unvaccinated calves shed virus in nasal swabs for 6 days and showed 10 days of viremia, while the vaccinated calves showed 1 and 2 days of viremia (at 5 and 6 dpi). Regarding the clinical signs, unvaccinated animals presented gastrointestinal symptoms, including diarrhea and hematochezia, after challenge with BVDV-2b, meaning a score of 10. In contrast, vaccinated calves showed scores of 0 and 1 due to a mild diarrhea in one of the two calves at 9 dpi.

This pilot study suggests that there would be an association between the VN Ab titer induced by the vaccine in guinea pigs and calves and the degree of protection against disease caused by BVDV infection. Further challenge experiments with at least 5 to 10 calves per group, including vaccines of different quality, are needed for a proper statistical analysis. But, this is an important proof of principle that the challenge model in calves is well standardized and represent a useful tool together with the guinea pig model to test the efficacy of vaccines to be registered in the local market (Figure 5.9).

5.6 Discussion and conclusions

The guinea pig model is an optimal tool to evaluate the quality of killed and monovalent and multivalent subunit vaccines, both oil- and aqueous-adjuvanted, to prevent BVDV.

The response pattern was similar in both animal species for both the kinetics of the Ab response and the magnitude of the Ab VN anti-BVDV response.

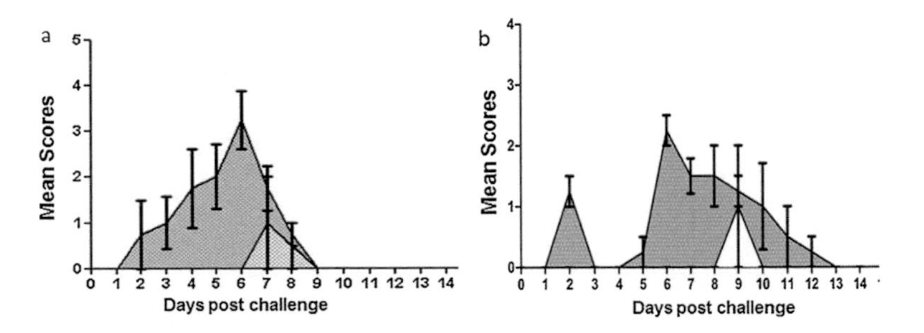

Figure 5.9 Clinical score of animals vaccinated with the subunit vaccine and then challenged with BVDV-2 (a) or BVDV-1b (b). Area under curve (AUC) in dark gray indicates the clinical score (CS) of unvaccinated animals (n = 4). (a) Challenge with BVDV-1b (98/204): one calf was fully protected, and the other (AUC in light gray) showed a CS = 3. (b) Challenge with BVDV-2b (98/124): One calf was fully protected, and the other (AUC in white) showed a CS = 2. (Adapted from Pecora et al. 2015).

The VN Ab response at 30 pvd in guinea pigs is adequate to perform the analysis at that single timepoint to compare with bovines sampled at 60 pvd. In both types of formulation, only the vaccines formulated with antigen concentrations in the order of 10^8 and 10^7 $TCID_{50}$/dose induced detectable levels of VN Ab, establishing the optimal time of sampling for the control and the range of immunogenicity of these vaccines.

The dose–response results allowed the estimation of split points to classify vaccines in guinea pigs and cattle. The concordance between the lab animal model and the target species was substantial to almost perfect for classifying the vaccines.

In addition, a calf model of BVDV infection and disease for two BVDVs was also standardized (Malacari et al. 2016, 2018) and showed good association between the VN Ab titer induced by a vaccine in guinea pigs and in seronegative calves and protection against infection and disease in calves (Pecora et al. 2015). It constitutes an excellent tool as a registration test for new vaccines to be launched on the market (Bellido et al. 2021).

The guinea pig model can be used for quality control of each batch of BVDV vaccine to be released to the market as well as for extension of expiration dates. It constitutes a practical tool both for vaccine companies and for the official animal health authorities to ensure the presence of effective products in the market.

The recommendation guideline of the guinea pig model for potency testing of BVDV vaccines was elaborated by the ad hoc group at PROSAIA

Foundation and submitted to the American Committee for Veterinary Medicines (CAMEVET) in 2016. A redaction commission was created. However, to finalize the validation, we decided to standardize a blocking ELISA to measure Ab in serum of bovines and guinea pigs. The assay is based on the E2 protein captured by a llama-derived nanobody, and serum samples are run in a ¼ dilution competing with a rabbit anti-E2 antiserum. The assay can be used to complement or replace the VN assay and was shown to be an excellent tool to test the vaccine in large field trials (Bellido et al. 2021). We have already estimated preliminary split points by ELISA for cattle (see Chapter 6), and the split point for guinea pigs and concordance analysis are in process.

Chapter 5 Annex IV

Potency test for BVDV vaccines in the guinea pig model

Guidelines elaborated by the PROSAIA biologic ad hoc group and presented to CAMEVET in 2016. The document remains in stage III of discussion by the redactor group of countries (Argentina, Brazil, Colombia, Chile, Uruguay, and the United States) and has not been voted yet due to the Covid-19 pandemic. More results are being generated by ELISA and will be added in order to improve the validation of the model.

Guinea pigs

At least six animals older than 30 days of age are used per vaccine. Animals' weight should be 400 grams ± 50 grams, and they must come from a controlled colony free of antibodies against BVDV. Males and females can be used, but each group must contain animals of the same gender. Animals must have a minimum period of adaptation of SEVEN (7) days after admission to the inoculation room.

Procedure

The guinea pigs are immunized with two doses of vaccine (within 21 days), subcutaneously, with a volume corresponding to one-fifth of the bovine dose. The animals are kept under control for a minimum of 30 days, and serum samples are taken at 9 days post booster (30 pvd). A group of non-vaccinated control guinea pigs (n = 2) should be included in the evaluation of the unknown vaccine (n = 6). Lastly, a group of guinea pigs vaccinated with a reference vaccine (n = 6) should be used in every assay. Thirty (30) pdv, the vaccinated animals and the controls are bled, and the serological control is performed by the VN assay.

Viral neutralization assay to determine Ab for BVDV

Reagents

Virus used: BVDV Singer reference strain (CP, genotype 1a). Diluted so as to contain 100 tissue culture infectious dose 50% ($TCID_{50}$).

The guide for vaccines against BVDV is based on correlations of neutralizing antibody titers against a strain of genotype 1a. However, this VN protocol can be used for any other cytopathic strain of BVDV (genotype 1b, 2, or 3).

Controls

Guinea pig positive control: Serum pool of five guinea pigs vaccinated with two doses of vaccine formulated with 10^7 $TCID_{50}$/ml of BVDV in oil adjuvant (Reference Vaccine) with a neutralizing Ab titer of 2.4–3.0.

Guinea pig negative control: normal guinea pig serum pool (pre-immunized guinea pig sera or unvaccinated controls).

Positive standard: immunized guinea pig serum with known VN Ab titer.

Negative standard: normal guinea pig serum.

Cell suspension: a cell suspension of MDBK line containing 200,000–250,000 cells/ml is used.

Sample inactivation: before being used in a VN assay, serum samples, including assay controls, must be heated in a water bath at 56 ± 3°C for 30 ± 5 minutes to inactivate the complement.

Preparation of working medium: MEM-E supplemented with 1% antibiotic solution (0.5% gentamicin sulfate, 0.7% streptomycin sulfate, 0.2% penicillin G sodium), and 2% gamma-irradiated bovine fetal serum (FBS).

Viral neutralization assay procedure with fixed virus–variable serum

1. 96-well culture plates are used. Place 60 μl/well of medium in all plates to be used.
2. Design of plate for sera tested: add 60 μl of the sample being tested, in quadruplicate. Start with a minimum dilution of 1/2, which summed to the volume of virus and cells, results in an initial dilution of 1/4. Include standards of known titer at random among the samples to be analyzed.
3. Design of control plate: positive and negative control sera are placed in the same way as the samples. To perform the cell control, add 60 μl of working medium per quadruplicate in 4 rows (16 wells in total). For the control of the 100 $TCID_{50}$ of virus, three 10-fold dilutions are carried out based on the working dilution. 100 μl of each of the fourfold dilutions prepared is added in quadruplicate (pure, 1/10, 1/100, and 1/1000).
4. Carry out twofold serial dilutions, transferring 60 μl, for all samples and control sera.
5. A toxicity control is carried out for each sample, adding 60 μl of medium in another plate.
6. Prepare the dilution of the working virus (100 $TCID_{50}$) in the working medium. Add 60 μl of the dilution of working virus to all plates, except

for the plate of toxicity controls, in the cells control, and in the 100 $TCID_{50}$ control.

7. Carry out three 10-fold dilutions of the working virus (pure, 1/10, 1/100, 1/1000) and add four replicates of each dilution to a separate plate. Incubate the plates (serum–virus blend) for 1 hour at 37°C in an atmosphere containing 5% CO_2.
8. Add 100 μl of the cell suspension containing 200,000–250,000 cells/ml per well to the serum–virus blend in all plates. Incubate plates at 37 ± 1°C in an atmosphere containing 5 ± 1% CO_2 for 48–72 h.

Reading and interpretation: after 48–72 h, reading is performed by inspection of monolayers in an optical microscope. Reading is by observation of viral cytopathic effect (CPE) typical of BVDV. Wells presenting a CPE typical of BVDV are considered positive. In toxicity controls, the monolayer must be observed to be the same as in the cell controls, free of CPE and free of toxic effect. The neutralizing titer of the analyzed serum is obtained by the quantity of protected replicates in the serial dilutions based on the Reed and Muench interpolation method. If a certain serum presents toxicity in the analyzed dilutions, the neutralizing antibodies titer will not be determined by this technique.

Assay acceptance (conformity): The assay is accepted when:

- Monolayers of cell controls are in good condition (confluent monolayers, light-refracting cells, with no morphological alterations, with no signs of contamination, and with no BVDV CPE).
- Viral suspension titer contains 100 $TCID_{50}$, with an acceptable range of 50–200 $TCID_{50}$.
- Positive control shows the expected titer ± one well.
- Negative control results are negative. An arbitrary value of 0.3 is assigned for calculation purposes

Reporting results of immunogenetic quality of BVDV vaccines tried/tested in guinea pigs and sera evaluated by VN

Test validation criteria

The guinea pig potency test will be valid when the VN Ab average titer obtained from the animals vaccinated with standard satisfactory vaccine results in the expected value and unvaccinated guinea pigs (controls) remain seronegative for VN Abs against BVDV throughout the experiment.

Calculations

All serum samples from the six animals immunized with the control vaccine will be evaluated. The FIVE (5) sera with the highest titer obtained (expressed in $\log_{10}$ of the inverse of the maximum dilution with neutralizing activity) are selected, and the arithmetic average is calculated. Results can be interpreted and used to classify a vaccine in the guinea pig model by VN only if the viral neutralization assay has been accepted and a minimum of five animals with VN Ab titer results were obtained to generate the mean Ab titer induced by the vaccine.

To validate the VN assay, sera of guinea pigs immunized with the reference vaccine must show a mean titer within the range established by a control chart that shows the mean value ± two standard deviations obtained from a minimum of five samples.

For the vaccine under evaluation, the mean neutralizing anti-BVDV Ab titer obtained by the Reed and Muench method of the best five immunized guinea pigs must be reported. Negative samples in the minimum serum dilution assayed (1/8) are expressed as an arbitrary titer of 0.3 for calculation purposes.

Approval criteria

If the $\log_{10}$ average of the VN Ab titers at 30 dpv in the vaccinated guinea pigs is greater than or equal to 1.37, the vaccine will be considered satisfactory.

Regional assay harmonization

There should be a panel of positive and negative control sera for BVDV antibodies as well as standard vaccines. These local reference reagents will be used to harmonize the results obtained by each laboratory adopting this control method. The standard vaccine will ensure the conformity of the guinea pig immunization test, while the serum panel may be used as a control of the recommended serological technique (VN) and for the standardization of alternative assays (ELISA).

6
Guinea Pig Model Application

6.1 Results of the application of the model in Argentina for IBR and rotavirus vaccines

Since 2008, the Argentinean animal health service SENASA has started to survey the local market, applying the guinea pig model developed by the National Agriculture Technology Institute (INTA) to control the potency of the combined vaccines applied in cattle for two strategic viral antigens, BoHV-1 and rotavirus. BoHV-1 causes infectious bovine rhinotracheitis (IBR) and is present in all respiratory and reproductive vaccines, while rotavirus is the viral antigen present in the vaccines to prevent calf scours.

For the implementation of the vaccine control, a facility to host the guinea pigs was prepared in SENASA (Figure 6.1a). From the vaccine batches produced every year, SENASA randomly selected batches to be tested. The vaccines were introduced into special containers and codified, losing their identity during the entire process of guinea pig vaccination and sampling in order to do a double-blind test (Figure 6.1b). Serum samples taken at 30 post-vaccination days (pvd) were submitted to INTA for serologic analysis. When the results were ready, all cans were opened in the presence of a representative of each vaccine company, and the vaccine classification was reported.

To conduct this quality control, two reference vaccines, one trivalent (IBR, bovine viral diarrhea virus [BVDV], and parainfluenza virus type 3 [PI-3]) (Figure 6.2) and the other for rotavirus (Figure 6.3), were formulated and included in each guinea pig experiment as positive control.

Regarding the vaccine testing for BoVH-1, the survey started in 2007, and the method was fully transferred to SENASA in 2011. In this initial stage, in 2011, a total of 173 vaccines were presented by the vaccine companies to be commercialized; 16% (28/173) were tested in guinea pigs, and 28.6% (8/28) were rejected (Figure 6.4a and b).

In 2012, the approval rate increased to 82.6%. Based on the robustness achieved in the validation of the model for IBR and the results of the evaluation of the tool from 2008 to 2011, SENASA issued a sanitary resolution

DOI: 10.1201/9781003373148-6

Figure 6.1 (a) SENASA facility to conduct the vaccine potency testing in guinea pigs. (b) Cans with vaccine codification to conduct a double-blind test.

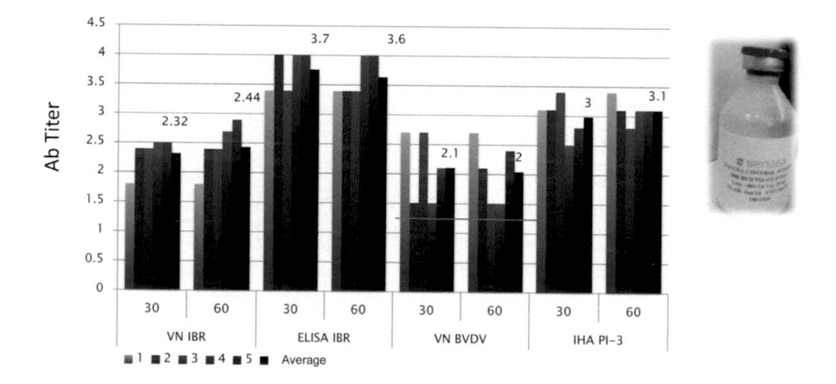

Figure 6.2 Trivalent reference vaccine to be included in each potency test experiment of respiratory or reproductive vaccines containing 10^7 TICD50/dose of IBR, BVDV, and PI-3. The graph depicts the Ab titer at 30 and 60 pvd in five guinea pigs and the average (black labeled bar) at 12 months post-elaboration, indicating that the vaccine remained satisfactory for the three antigens.

(Resol. 598.12) and adopted the INTA guinea pig model and the associated enzyme-linked immunosorbent assay (ELISA) tests as the official method to control the potency of the vaccines to be commercialized in Argentina. In 2013, after the implementation of the official control (the vaccine batches not passing the guinea pig test were not approved to be released into the market), the percentage of controlled batches was 37%, and the approval rate reached 96.6% (Figure 6.4b). In 2014 and 2015, a total of 172 and 128 vaccines containing IBR were presented, 39% and 64% were tested, and the approval rate was 76% and 98.5%, respectively, indicating that the

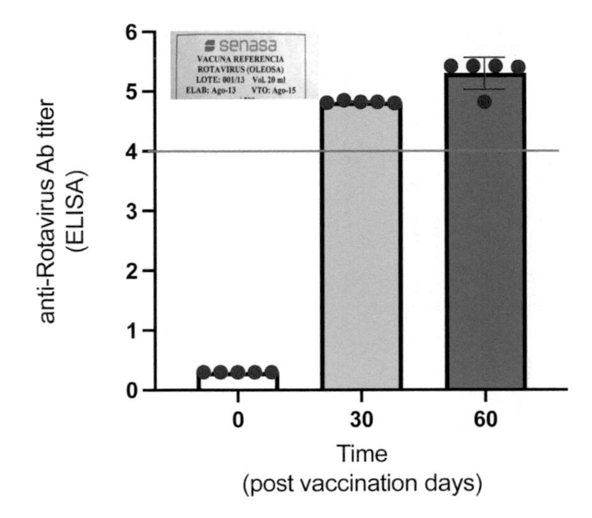

Figure 6.3 Rotavirus reference vaccine formulated with 1×10^7 focus forming assay units (FFU)/dose of rotavirus G6P[5] and G10P[11] to be included in each potency test experiment on vaccines to prevent neonatal calf diarrhea. The graph depicts the Ab titer at 30 and 60 pvd in five guinea pigs and the average (light blue bar) classified as satisfactory at 21 months post-elaboration.

implementation of the control had clearly improved the IBR vaccine quality over time (Figure 6.4c).

Regarding rotavirus, the evaluation of the model by SENASA started in 2008, and the guinea pig model was fully validated in 2012, together with the ELISAs to measure IgG antibodies in guinea pigs and IgG1 antibodies in cattle, given that this isotype is the one that is transferred from the dam to the calf via colostrum and milk to achieve protection against neonatal diarrhea (Parreño et al. 2004). In general, the rotavirus vaccines showed good quality (Figure 6.5). In 2013, when the sanitary resolution was implemented, and rotavirus vaccines were also included in the punitive control, a total of 14 vaccines were presented, 5 were tested, and 40% (2/5) were approved. In subsequent years, the approval rates improved, reaching more than 80% approval (Table 6.1).

From 2015 onward, SENASA took over control, doing the immunization and the serology for the potency testing of IBR and rotavirus vaccines, while INTA has provided both ELISA kits since then. In 2016, a total of 138 vaccine batches were produced containing IBR or rotavirus; 50% (69/138) were tested in the guinea pig model, and 95% (66/69) were approved, indicating that the success of the program translated into a substantial improvement in the quality of the vaccines present in the market.

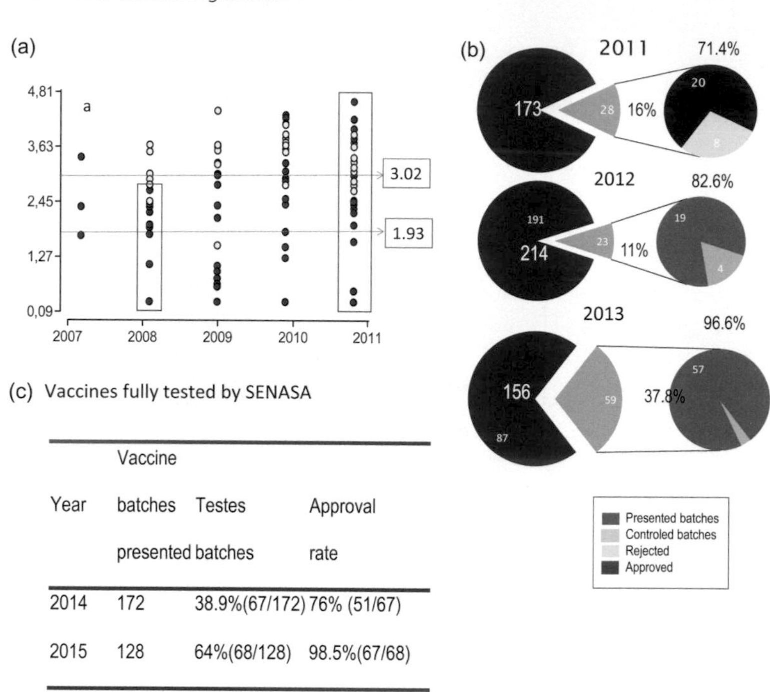

(c) Vaccines fully tested by SENASA

Year	Vaccine batches presented	Testes batches	Approval rate
2014	172	38.9%(67/172)	76% (51/67)
2015	128	64%(68/128)	98.5%(67/68)

Figure 6.4 Respiratory and reproductive vaccines tested using the INTA guinea pig model for IBR. (a) Vaccines tested from 2007 to 2011, guinea pig immunization at SENASA, serology at INTA. Blue dots represent the mean Ab titer of five guinea pigs tested with each vaccine, yellow dots reference vaccines included in each assay. (b) Percentage of vaccines presented, tested, approved, and rejected from 2011 to 2015. In 2013, the sanitary resolution 598.12 entered into force, and vaccines not passing the cut-off were seized and destroyed. (c) In 2014 and 2015, the vaccines were fully tested by SENASA with ELISA kits provided by INTA.

6.2 Survey of PI-3 vaccines in Argentina

Regarding PI-3, the model was validated in 2008, together with IBR. The market was surveyed from 2008 to 2014 using the hemagglutination inhibition assay (HIA) (Figure 6.6). The overall approval rate was 75%; however, it is interesting to see that the number of rejected vaccines increased over time. The explanation was that different vaccines were tested, and once a formulation was rejected, every batch of that vaccine was tested. As mentioned in Chapter 3, an ELISA test has been recently validated (Maidana et al. 2021), and it will be very important to provide this to the animal health services with the model and a more versatile serology test to implement the punitive control for this antigen.

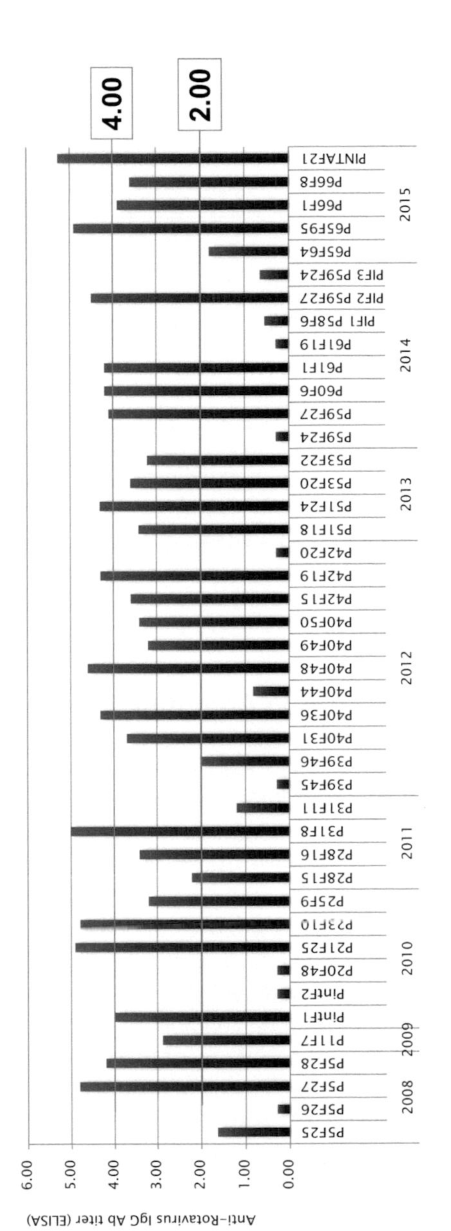

Figure 6.5 Rotavirus vaccines tested using the INTA guinea pig model from 2008 to 2015. In 2013, the sanitary resolution 598.12 entered into force, and vaccines not passing the cut-off were not released to the market. The bars represent the mean Ab titer of five guinea pigs immunized with each vaccine. The red lines indicate the split point from vaccine classification 1.96 = intermediate and 4 = satisfactory.

Table 6.1 Rotavirus Vaccines Tested Using the Guinea Pig Model

Year	Vaccine batches presented	Tested batches	Approval rate
2013	14	35.7% (5/14)	40% (2/5)
2014	17	71% (12/17)	92% (11/12)
2015	11	64% (7/11)	86% (6/7)

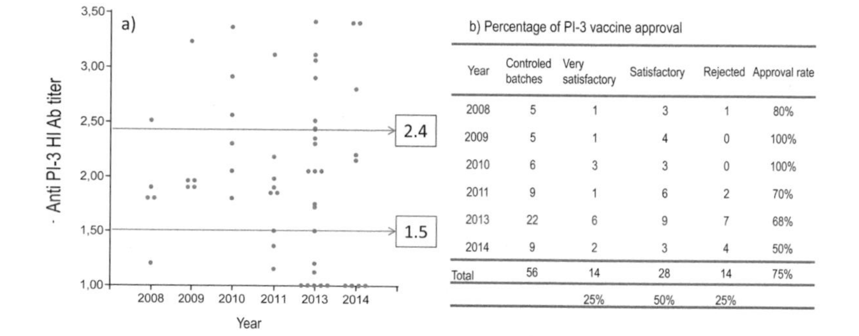

Year	Controled batches	Very satisfactory	Satisfactory	Rejected	Approval rate
2008	5	1	3	1	80%
2009	5	1	4	0	100%
2010	6	3	3	0	100%
2011	9	1	6	2	70%
2013	22	6	9	7	68%
2014	9	2	3	4	50%
Total	56	14	28	14	75%
		25%	50%	25%	

Figure 6.6 Respiratory vaccines containing PI-3 tested using the INTA guinea pig model from 2008 to 2013. (a) The red dots represent the mean HI Ab titer of five guinea pigs immunized with each vaccine. The blue lines indicate the split point from vaccine classification 1.5 = satisfactory and 2.4 = very satisfactory. (b) Table detailing the approval rate over time.

6.3 Survey of BVDV vaccines in Argentina

Finally, the model for BVDV has been evaluated by SENASA since 2013. Guinea pigs were immunized in SENASA, and the serum samples were submitted to INTA to run the viral neutralization (VN) assay. A total of 56 vaccines were evaluated between 2013 and 2014; 8 vaccines induced detectable VN Ab titers, but only 3 passed the cut-off selected to approve the vaccines (Figure 6.7a). From 2015 to 2018, the reference vaccine was included in the tests (red bars). In this second stage, a total of 30 vaccines were tested; 63% (19/30) were rejected, while 37% (11/30) were approved. This was the antigen with the highest percentage of rejection. For this reason, INTA decided to accelerate the development of a targeted subunit vaccine (VEDEVAX), which was launched on the market in 2018 with the aim of providing the livestock industry with a very good vaccine for BVDV based on an innovative technology (Bellido et al. 2021). The other companies also made efforts to improve their formulations (Figure 6.7b).

Understanding that an ELISA assay was a very important tool to develop together with the guinea pig model and the subunit vaccine for BVDV, in order to have the VN as a gold standard assay and a second, more versatile

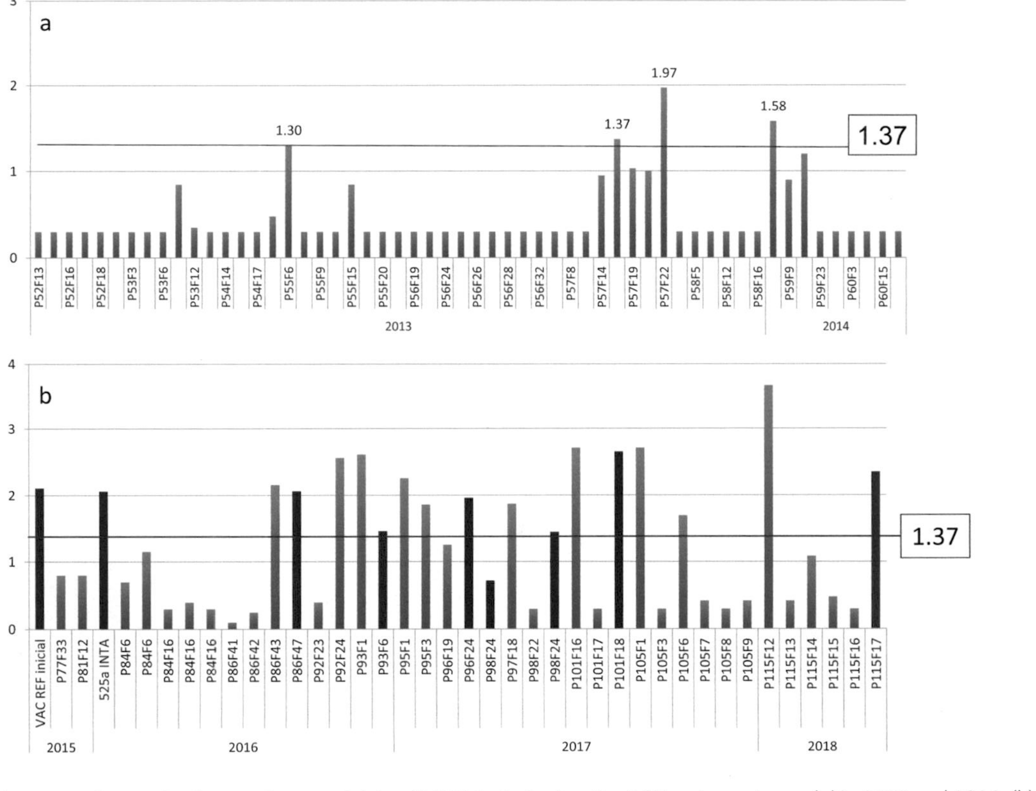

Figure 6.7 (a) Respiratory and reproductive vaccines containing BVDV, tested using the INTA guinea pig model in 2013 and 2014. (b) From 2015 to 2018. The blue bars represent the mean VN Ab titer of five guinea pigs immunized with each vaccine. The red bars indicate the reference vaccine. The red line indicates the cut-off point selected from vaccine classification 1.37 = satisfactory.

and quantitative assay to run a large number of samples, we developed a competition ELISA.

Briefly, the ELISA consists of plates sensibilized with a llama-derived nanobody directed to the E2 protein of BVDV 1a to capture the protein expressed in baculovirus. E2 was chosen because it is the neutralizing antigen of BVDV. Bovine or guinea pig serum samples are then added in a unique ¼ dilution followed by a rabbit polyclonal antiserum to E2 and a peroxidase-labeled anti-rabbit IgG. The ELISA Ab titers are expressed as a percentage of displacement (PD%) of the positive rabbit anti-E2 hyperimmune serum, which is considered the 100% signal (Figure 6.8a). The cut-off of the assay was established as 10% PD. Tentative split points for vaccine classification were estimated in bovines (Figure 6.8b). The ELISA Ab titers showed optimal correlation with the VN Ab titers of a set of bovine samples tested, including negative and positive animals from infection and/or vaccination (Figure 6.8c).

The ELISA was used to test vaccines in field trials conducted in beef herds. We can see, for example, that the APCH-E2t subunit vaccine was able to induce high Ab titers, boosting the Ab response in already primed animals (seropositive prior due to vaccination or infection) and inducing very strong responses in naïve (seronegative) cattle after two doses administered at 0 and 30 pvd. In contrast, an aqueous commercial reproductive vaccine was able to boost the Ab responses in seropositive animals reaching titers around 35% by 60 pvd but only produced a slight increase from 0% to ~12% in naïve animals by 120 pvd (Figure 6.9) (Bellido et al. 2021).

6.4 INTA services

Table 6.2 summarizes the split points for vaccine classification in guinea pigs and bovines for the viral antigen studied by the different serological techniques used.

Applying the split points detailed in Table 6.2, INTA continues to provide the service for vaccine potency testing in guinea pigs to the vaccine companies, which annually send vaccines and sera for evaluation. Likewise, companies can acquire ELISA kits to carry out their controls internally. The validation of the guinea pig model and the corresponding ELISAs to evaluate coronavirus vaccines is currently in progress; we have already obtained the split points for guinea pigs, and we are conducting the dose–response experiments in bovines.

Other services that INTA also provides are field trials to study vaccine performance and the challenge in calves for IBR, rotavirus, and BVDV for efficacy

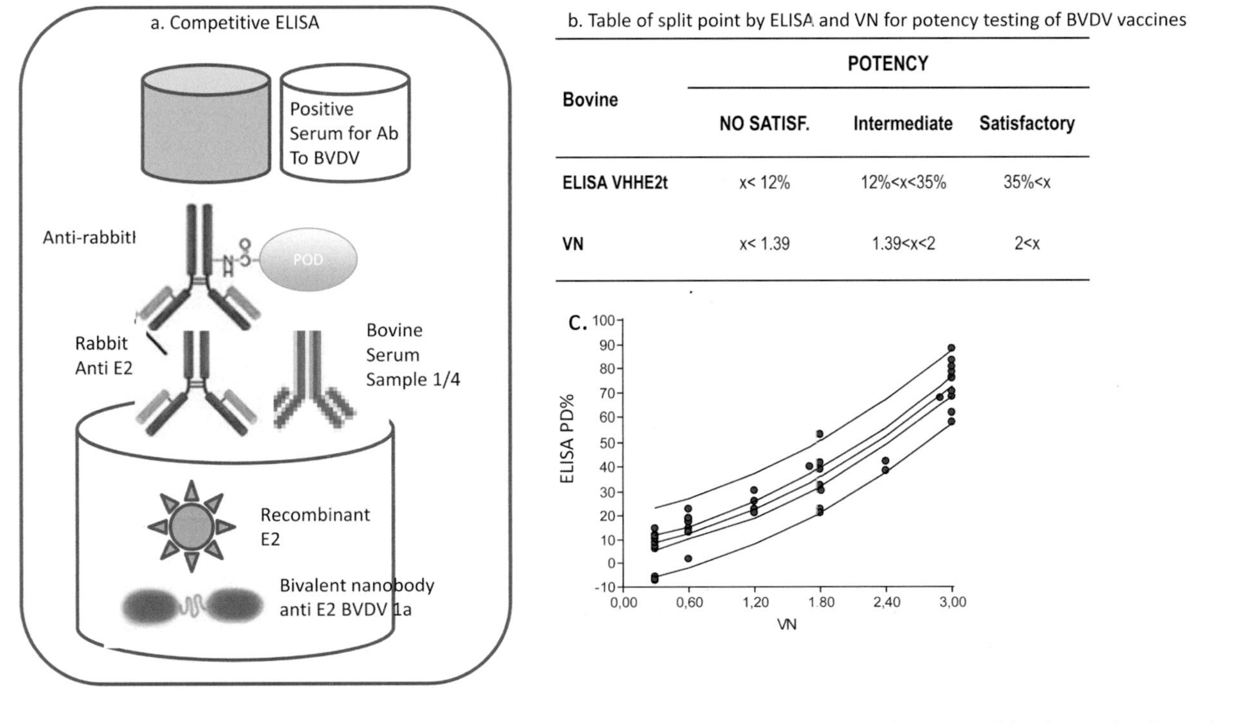

b. Table of split point by ELISA and VN for potency testing of BVDV vaccines

Bovine	POTENCY		
	NO SATISF.	Intermediate	Satisfactory
ELISA VHHE2t	x< 12%	12%<x<35%	35%<x
VN	x< 1.39	1.39<x<2	2<x

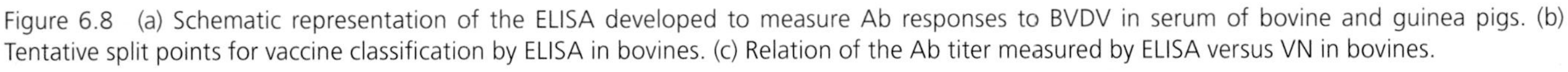

Figure 6.8 (a) Schematic representation of the ELISA developed to measure Ab responses to BVDV in serum of bovine and guinea pigs. (b) Tentative split points for vaccine classification by ELISA in bovines. (c) Relation of the Ab titer measured by ELISA versus VN in bovines.

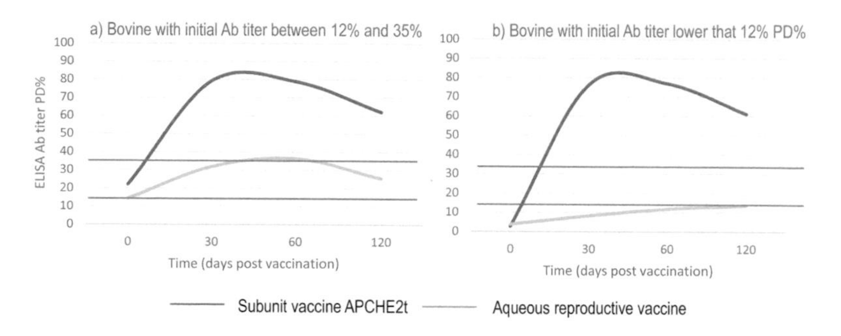

Figure 6.9 Kinetics of anti-BVDV Ab titers measured by ELISA in beef cattle vaccinated with the APCH-E2t subunit targeted vaccine or a reproductive aqueous vaccine (2 doses at 0 and 30 pvd). (a) Responses in seropositive animals. (b) Ab Responses in naïve animals, negative for Ab to BVDV by ELISA and VN.

evaluation of new vaccines to be registered and launched into the local market.

Since 2010, in parallel with the implementation of the control, a cycle of training courses was started in two formats: an annual course that is taught at INTA, which is especially aimed at postgraduate veterinary students; and tailor-made courses taught in each vaccine company. This modality has been carried out successfully in most of the vaccine companies from Argentina and Uruguay. We also started a scientific collaboration with the University of Sao Paulo, Brazil. A first course was given in 2019, and the ELISA techniques for testing IBR, rotavirus A (RVA), and BVDV were transferred. Finally, we provided reference material for a Spanish company and gave training to professionals from a vaccine company from Turkey in 2021.

6.5 Guideline recommendations elaborated in PROSAIA Foundation and elevated to the CAMEVET-OIE

In 2011, the PROSAIA Foundation created an ad hoc group to write recommendation guidelines for biologicals. The group included specialists from SENASA and INTA, representatives of the vaccine company committees (CAPROVE and CLAMEVET), as well as specialists from the academic sector (the National Scientific and Technical Research Council [CONICET] and universities). This forum developed a set of guidelines for the control of vaccines using the guinea pig model that were submitted for consideration to the American Committee for Veterinary Medicines (CAMEVET), the World Organisation for Animal Health (OIE) regional office for the Americas. To date, three guides have been voted unanimously as CAMEVET documents. The IBR guideline was approved in the meeting that was held in Panama in

Table 6.2 Split Points for IBR, Rotavirus, PI-3, and BVDV Vaccine Potency Testing

			Vaccine potency		
Viral antigen	**Species**	**Technique**	**Not satisfactory**	**Satisfactory**	**Very satisfactory**
IBR	Guinea pig	ELISA	<1.93	$1.93 \leq X < 3.02$	$x \geq 3.02$
		VN	<1.32	$1.32 \leq X < 2.05$	$x \geq 2.05$
	Bovine	ELISA	<1.69	$1.69 \leq X < 2.72$	$x \geq 2.72$
		VN	<1.27	$1.27 \leq X < 1.96$	$x \geq 1.96$
RVA	Guinea pig	ELISA	<1.96	$1.96 \leq X < 4.00$	$4.00 \leq x \leq 4.75$
	Bovine final Ab titer 60 pvd	ELISA	<3.53	$3.53 \leq X < 3.70$	$x \geq 3.70$
Guinea PI-3	Guinea pig	HIA	<1.5	$1.5 \leq X < 2.4$	$x \geq 2.4$
	Bovine final Ab titer 60 pvd	HIA	<2.8	$2.8 \leq X < 3.1$	$x \geq 3.1$
BVDV	Guinea pig	VN	<1.37	$1.37 \leq X < 2.04$	$x \geq 2.04$
	Bovine final Ab titer	VN	<1.39	$1.39 \leq X < 2.0$	$x \geq 2.08$
		ELISA	<12	$12 \leq X < 35$	$x \geq 35$

X denotes Bovine; x denotes Guinea pig.

2013. The guidelines for PI-3 and RVA were voted in Guatemala in 2015. The guideline for BVDV was submitted in 2016 in Mexico, and it is still under evaluation.

6.6 Beneficiaries of the implementation of the INTA guinea pig model and harmonized vaccine potency testing

⟹Beef and dairy farmers (end users of vaccines)

⟹The veterinary vaccine companies (main private clients)

⟹The animal health services of each country (main public client and entity adopting the technology for the official vaccine quality control)

⟹CAMEVET-OIE and other International agencies regulating vaccine trade.

6.7 Strategic importance

⟹To propose a method for potency testing of combined vaccines including viral antigens in their formulations

⟹To strengthen the capacities of the countries to control the quality of national and imported veterinary vaccines

⟹To facilitate the registration of vaccines in the scope of Mercosur and other regions

⟹To have a regional reference laboratory that develops, validates, and makes available to users, tools and reagents that allow the control of their biologicals (ELISA kits, viral strains, reference vaccines, serum panels, etc.)

⟹To provide a training system and continuous quality control: customized training courses and interlaboratory tests

References

9CFR 113.215, BVDV. 2014. 'Bovine Virus Diarrhea Vaccine, Killed Virus', in.

9CFR 113.216, IBR. 2014. 'Bovine Rhinotracheitis Vaccine, Killed Virus', in.

9CFR 114.309, PI-3. 2014. '113.309 Bovine Parainfluenza3 Vaccine', in.

Aguirre, I. M., M. P. Quezada, and M. O. Celedon. 2014. 'Antigenic variability in bovine viral diarrhea virus (BVDV) isolates from alpaca (Vicugna pacos), llama (Lama glama) and bovines in Chile', *Vet Microbiol*, 168: 324–30.

Aida, V., V. C. Pliasas, P. J. Neasham, J. F. North, K. L. McWhorter, S. R. Glover, and C. S. Kyriakis. 2021. 'Novel vaccine technologies in veterinary medicine: A Herald to human medicine vaccines', *Front Vet Sci*, 8: 654289.

Akkermans, A., J. M. Chapsal, E. M. Coccia, H. Depraetere, J. F. Dierick, P. Duangkhae, S. Goel, M. Halder, C. Hendriksen, R. Levis, K. Pinyosukhee, D. Pullirsch, G. Sanyal, L. Shi, R. Sitrin, D. Smith, P. Stickings, E. Terao, S. Uhlrich, L. Viviani, and J. Webster. 2020. 'Animal testing for vaccines. Implementing replacement, reduction and refinement: challenges and priorities', *Biologicals*, 68: 92–107.

Alocilla, O. A., and G. Monti. 2022. 'Bovine Viral Diarrhea Virus within and herd prevalence on pasture-based dairy systems, in southern Chile dairy farms', *Prev Vet Med*, 198: 105533.

Badaracco, A., L. Garaicoechea, J. Matthijnssens, E. Louge Uriarte, A. Odeon, G. Bilbao, F. Fernandez, G. I. Parra, and V. Parreño. 2013. 'Phylogenetic analyses of typical bovine rotavirus genotypes G6, G10, P[5] and P[11] circulating in Argentinean beef and dairy herds', *Infect Genet Evol*, 18: 18–30.

Badaracco, A., L. Garaicoechea, D. Rodriguez, E. L. Uriarte, A. Odeon, G. Bilbao, R. Galarza, A. Abdala, F. Fernandez, and V. Parreño. 2012. 'Bovine rotavirus strains circulating in beef and dairy herds in Argentina from 2004 to 2010', *Vet Microbiol*, 158: 394–9.

Bahnemann, H. G. 1990. 'Inactivation of viral antigens for vaccine preparation with particular reference to the application of binary ethylenimine', *Vaccine*, 8: 299–303.

Balint, A., C. Baule, V. Palfi, and S. Belak. 2005. 'Retrospective genome analysis of a live vaccine strain of bovine viral diarrhea virus', *Vet Res*, 36: 89–99.

Bartels, C. J., M. Holzhauer, R. Jorritsma, W. A. Swart, and T. J. Lam. 2010. 'Prevalence, prediction and risk factors of enteropathogens in normal and non-normal faeces of young Dutch dairy calves', *Prev Vet Med*, 93: 162–9.

Bauermann, F. V., and J. F. Ridpath. 2015. 'HoBi-like viruses--the typical 'atypical bovine pestivirus'', *Anim Health Res Rev*, 16: 64–9.

Bellido, D., J. Baztarrica, L. Rocha, A. Pecora, M. Acosta, J. M. Escribano, V. Parreño, and A. Wigdorovitz. 2021. 'A novel MHC-II targeted BVDV subunit vaccine induces a neutralizing immunological response in guinea pigs and cattle', *Transbound Emerg Dis*, 68: 3474–81.

Benoit, A., J. Beran, and J. M. Devaster. 2015. 'Hemagglutination inhibition antibody titers as a correlate of protection against seasonal A/H3N2 influenza disease', *Open Forum Infect Dis*, 2: ofv067.

Bertoni, E. A., M. Bok, C. Vega, G. M. Martinez, R. Cimino, and V. Parreño. 2021. 'Influence of individual or group housing of newborn calves on rotavirus and coronavirus infection during the first 2 months of life', *Trop Anim Health Prod*, 53: 62.

Bertoni, E. A., M. Aduriz, M. Bok, C. Vega, L. Saif, D. Aguirre, R. O. Cimino, S. Mino, and V. Parreño. 2020. 'First report of group A rotavirus and bovine coronavirus associated with neonatal calf diarrhea in the northwest of Argentina', *Trop Anim Health Prod*, 52: 2761–68.

Birck, M. M., P. Tveden-Nyborg, M. M. Lindblad, and J. Lykkesfeldt. 2014. 'Non-terminal blood sampling techniques in guinea pigs', *J Vis Exp*: e51982.

Blanchard, P. C. 2012. 'Diagnostics of dairy and beef cattle diarrhea', *Vet Clin North Am Food Anim Pract*, 28: 443–64.

Bolin, S. R. 1995. 'The pathogenesis of mucosal disease', *Vet Clin North Am Food Anim Pract*, 11: 489–500.

Bretz, F., M. Posch, E. Glimm, F. Klinglmueller, W. Maurer, and K. Rohmeyer. 2011. 'Graphical approaches for multiple comparison procedures using weighted Bonferroni, Simes, or parametric tests', *Biom J*, 53: 894–913.

Brock, J., M. Lange, J. A. Tratalos, S. J. More, M. Guelbenzu-Gonzalo, D. A. Graham, and H. H. Thulke. 2021. 'A large-scale epidemiological model of BoHV-1 spread in the Irish cattle population to support decision-making in conformity with the European Animal Health Law', *Prev Vet Med*, 192: 105375.

Brunauer, M., F. F. Roch, and B. Conrady. 2021. 'Prevalence of worldwide neonatal calf diarrhoea caused by bovine rotavirus in combination with bovine coronavirus, Escherichia coli K99 and Cryptosporidium spp.: A meta-analysis', *Animals*, 11: 1014.

CAMEVET. https://rr-americas.woah.org/en/projects/camevet/camevet-seminars/.

Chamorro, M. F., and R. A. Palomares. 2020. 'Bovine respiratory disease vaccination against viral pathogens: Modified-live versus inactivated antigen vaccines, intranasal versus parenteral, what is the evidence?', *Vet Clin North Am Food Anim Pract*, 36: 461–72.

Chamorro, M. F., P. H. Walz, T. Passler, R. Palomares, B. W. Newcomer, K. P. Riddell, J. Gard, Y. Zhang, and P. Galik. 2016. 'Efficacy of four commercially available multivalent modified-live virus vaccines against clinical disease, viremia, and viral shedding in early-weaned beef calves exposed simultaneously to cattle persistently infected with bovine viral diarrhea virus and cattle acutely infected with bovine herpesvirus 1', *Am J Vet Res*, 77: 88–97.

Cho, Y. I., and K. J. Yoon. 2014. 'An overview of calf diarrhea - infectious etiology, diagnosis, and intervention', *J Vet Sci*, 15: 1–17.

Cho, Y. I., J. I. Han, C. Wang, V. Cooper, K. Schwartz, T. Engelken, and K. J. Yoon. 2013. 'Case-control study of microbiological etiology associated with calf diarrhea', *Vet Microbiol*, 166: 375–85.

Di Rienzo, J., J. Guzman, and F. Casanoves. 2002 'Amultiple comparison method based on the distribution of the root node distance of a binary tree obtained by averagelinkage of the matrix of Euclidean distances between treatment means', *JABES*, 7: 1–14.

Donoso, A., F. Inostroza, M. Celedon, and J. Pizarro-Lucero. 2018. 'Genetic diversity of bovine viral diarrhea virus from cattle in Chile between 2003 and 2007', *BMC Vet Res*, 14: 314.

Dubovi, E. J., Y. T. Grohn, M. A. Brunner, and J. A. Hertl. 2000. 'Response to modified live and killed multivalent viral vaccine in regularly vaccinated, fresh dairy cows', *Vet Ther*, 1: 49–58.

Ellis, J. A. 2010. 'Bovine parainfluenza-3 virus', *Vet Clin North Am Food Anim Pract*, 26: 575–93.

EMEA/140/97. 1997. 'Position paper on compliance of veterinary vaccines with veterinary vaccine monographs of the European pharmacopoeia', in edited by CVMP VMEU.

EMEA/P038/97. 1998. 'Position paper on batch potency testing of immunological veterinary medical products', in edited by CVMP/IWP VMEU.

Extension, University of Minnesota. 'Cattle vaccine basics', https://extension.umn.edu/beef-cow-calf/cattle-vaccine-basics#revaccination-and-boostering--2068111.

Flores, E. F., J. F. Cargnelutti, F. L. Monteiro, F. V. Bauermann, J. F. Ridpath, and R. Weiblen. 2018. 'A genetic profile of bovine pestiviruses circulating in Brazil (1998–2018)', *Anim Health Res Rev*, 19: 134–41.

Foster, D. M., and G. W. Smith. 2009. 'Pathophysiology of diarrhea in calves', *Vet Clin North Am Food Anim Pract*, 25: 13–36, xi.

Fulton, R. W. 2015. 'Impact of species and subgenotypes of bovine viral diarrhea virus on control by vaccination', *Anim Health Res Rev*, 16: 40–54.

———. 2020. 'Viruses in Bovine Respiratory Disease in North America: Knowledge Advances Using Genomic Testing', *Vet Clin North Am Food Anim Pract*, 36: 321–32.

Geletu, U. S., M. A. Usmael, and F. D. Bari. 2021. 'Rotavirus in Calves and Its Zoonotic Importance', *Vet Med Int*, 2021: 6639701.

Gomez-Romero, N., F. J. Basurto-Alcantara, A. Verdugo-Rodriguez, F. V. Bauermann, and J. F. Ridpath. 2017. 'Genetic diversity of bovine viral diarrhea virus in cattle from Mexico', *J Vet Diagn Invest*, 29: 362–65.

Gomez-Romero, N., J. F. Ridpath, F. J. Basurto-Alcantara, and A. Verdugo-Rodriguez. 2021. 'Bovine viral diarrhea virus in cattle from Mexico: Current status', *Front Vet Sci*, 8: 673577.

Gomez-Romero, N., L. Velazquez-Salinas, J. F. Ridpath, A. Verdugo-Rodriguez, and F. J. Basurto-Alcantara. 2021. 'Detection and genotyping of bovine viral diarrhea virus found contaminating commercial veterinary vaccines, cell lines, and fetal bovine serum lots originating in Mexico', *Arch Virol*, 166: 1999–2003.

Gonda, M. G., X. Fang, G. A. Perry, and C. Maltecca. 2012. 'Measuring bovine viral diarrhea virus vaccine response: using a commercially available ELISA as a surrogate for serum neutralization assays', *Vaccine*, 30: 6559–63.

González Altamiranda, E. A., G. G. Kaiser, N. Weber, M. R. Leunda, A. Pecora, D Malacari, and Odeón. A. C. 2012. 'Clinical and reproductive consequences of using BVDV-contaminated semen in artificial insemination in a beef herd in Argentina', *Animal Reproduction Science*, 133: 146–52.

Griebel, P. J. 2015. 'BVDV vaccination in North America: risks versus benefits', *Anim Health Res Rev*, 16: 27–32.

Grissett, G. P., B. J. White, and R. L. Larson. 2015. 'Structured literature review of responses of cattle to viral and bacterial pathogens causing bovine respiratory disease complex', *J Vet Intern Med*, 29: 770–80.

Haanes, E. J., P. Guimond, and R. Wardley. 1997. 'The bovine parainfluenza virus type-3 (BPIV-3) hemagglutinin/neuraminidase glycoprotein expressed in baculovirus protects calves against experimental BPIV-3 challenge', *Vaccine*, 15: 730–8.

Halder, M., C. Hendriksen, K. Cussler, and M. Balls. 2002. 'ECVAM's contributions to the implementation of the Three Rs in the production and quality control of biologicals', *Altern Lab Anim*, 30: 93–108.

Harada, T., N. Ariyoshi, H. Shimura, Y. Sato, I. Yokoyama, K. Takahashi, S. Yamagata, M. Imamaki, Y. Kobayashi, I. Ishii, M. Miyazaki, and M. Kitada. 2010. 'Application of Akaike information criterion to evaluate warfarin dosing algorithm', *Thromb Res*, 126: 183–90.

Hendriksen, C. F. 1999. 'Validation of tests methods in the quality control of biologicals', *Dev Biol Stand*, 101: 217–21.

InfoStat, Grupo. 2008. 'InfoStat, versión 2008. Manual del Usuario', in edited by Universidad Nacional de Córdoba. *Primera Edición FCA, Editorial Brujas Argentina.*

Iscaro, C., V. Cambiotti, S. Petrini, and F. Feliziani. 2021. 'Control programs for infectious bovine rhinotracheitis (IBR) in European countries: an overview', *Anim Health Res Rev*, 22: 136–46.

Jungbäck, C. 2011. 'Potency testing of veterinary vaccines for animals: The way from in vivo to in vitro', in *Developments in Biologicals*, edited by International Alliance for Biological Standardization (IABS). Karger, 177. Switzerland. Worldwide distribution by S. Karger AG: International Alliance for Biological Standardization (IABS).

Kampa, J., K. Stahl, L. H. Renstrom, and S. Alenius. 2007. 'Evaluation of a commercial Erns-capture ELISA for detection of BVDV in routine diagnostic cattle serum samples', *Acta Vet Scand*, 49: 7.

Lanyon, S. R., F. I. Hill, M. P. Reichel, and J. Brownlie. 2014. 'Bovine viral diarrhoea: pathogenesis and diagnosis', *Vet J*, 199: 201–9.

Lemon, S. C., J. Roy, M. A. Clark, P. D. Friedmann, and W. Rakowski. 2003. 'Classification and regression tree analysis in public health: methodological review and comparison with logistic regression', *Ann Behav Med*, 26: 172–81.

Little, R. J., and T. Raghunathan. 1999. 'On summary measures analysis of the linear mixed effects model for repeated measures when data are not missing completely at random', *Stat Med*, 18: 2465–78.

Loy, J. D., K. A. Clothier, and G. Maier. 2021. 'Component causes of infectious bovine keratoconjunctivitis-non-moraxella organisms in the epidemiology of infectious bovine keratoconjunctivitis', *Vet Clin North Am Food Anim Pract*, 37: 295–308.

Maidana, S. S., M. Odeon, C. Ferrecio, N. Grazziotto, E. Pisano, I. Alvarez, L. Rocha, V. Parreño, and A. Romera. 2021. 'Development and validation of a bovine parainfluenza virus type 3 indirect ELISA', *Int J Environ Agri Biotechnol*, 6: 196–205.

Maidana, S. S., P. M. Lomonaco, G. Combessies, M. I. Craig, J. Diodati, D. Rodriguez, V. Parreño, O. Zabal, J. L. Konrad, G. Crudelli, A. Mauroy, E. Thiry, and S. A. Romera. 2012. 'Isolation and characterization of bovine parainfluenza virus type 3 from water buffaloes (Bubalus bubalis) in Argentina', *BMC Vet Res*, 8: 83.

Makoschey, B., and A. C. Berge. 2021. 'Review on bovine respiratory syncytial virus and bovine parainfluenza: usual suspects in bovine respiratory disease: a narrative review', *BMC Vet Res*, 17: 261.

Malacari, D. A., A. Pecora, M. S. Perez Aguirreburualde, N. P. Cardoso, A. C. Odeon, and A. V. Capozzo. 2018. 'In vitro and in vivo characterization of a typical and a high pathogenic bovine viral diarrhea virus type II strains', *Front Vet Sci*, 5: 75.

Malacari, D. A., A. Pecora, A. V. Capozzo, and A. C. Odeón. 2016. 'Guía para la crianza y mantenimiento de terneros privados de calostro para su utilización como modelo animal', in edited by Ediciones INTA, 1689–99. Buenos Aires, Argentina: Instituto Nacional de Tecnologia Agropecuaria (INTA).

Maresca, C., E. Scoccia, A. Dettori, A. Felici, R. Guarcini, S. Petrini, A. Quaglia, and G. Filippini. 2018. 'National surveillance plan for infectious bovine rhinotracheitis (IBR) in autochthonous Italian cattle breeds: Results of first year of activity', *Vet Microbiol*, 219: 150–53.

Matthijnssens, J., M. Ciarlet, S. M. McDonald, H. Attoui, K. Banyai, J. R. Brister, J. Buesa, M. D. Esona, M. K. Estes, J. R. Gentsch, M. Iturriza-Gomara, R. Johne, C. D. Kirkwood, V. Martella, P. P. Mertens, O. Nakagomi, V. Parreño, M. Rahman, F. M. Ruggeri, L. J. Saif, N. Santos, A. Steyer, K. Taniguchi, J. T. Patton, U. Desselberger, and M. Van Ranst. 2011. 'Uniformity of rotavirus strain nomenclature proposed by the Rotavirus Classification Working Group (RCWG)', *Arch Virol*, 156: 1397–413.

Maya, L., M. Macias-Rioseco, C. Silveira, F. Giannitti, M. Castells, M. Salvo, R. Rivero, J. Cristina, E. Gianneechini, R. Puentes, E. F. Flores, F. Riet-Correa, and R. Colina. 2020. 'An extensive field study reveals the circulation of new genetic variants of subtype 1a of bovine viral diarrhea virus in Uruguay', *Arch Virol*, 165: 145–56.

Moore, D. P., C. M. Campero, A. C. Odeon, J. C. Bardon, P. Silva-Paulo, F. A. Paolicchi, and A. L. Cipolla. 2003. 'Humoral immune response to infectious agents in aborted bovine fetuses in Argentina', *Rev Argent Microbiol*, 35: 143–8.

Muratore, E., L. Bertolotti, C. Nogarol, C. Caruso, L. Lucchese, B. Iotti, D. Ariello, A. Moresco, L. Masoero, S. Nardelli, and S. Rosati. 2017. 'Surveillance of infectious bovine Rhinotracheitis in marker-vaccinated dairy herds: Application of a recombinant gE ELISA on bulk milk samples', *Vet Immunol Immunopathol*, 185: 1–6.

Murphy, B. R., S. L. Hall, A. B. Kulkarni, J. E. Crowe Jr., P. L. Collins, M. Connors, R. A. Karron, and R. M. Chanock. 1994. 'An update on approaches to the development of respiratory syncytial virus (RSV) and parainfluenza virus type 3 (PIV3) vaccines', *Virus Res*, 32: 13–36.

Odeón, A. C., E. J. A. Späth, E. J. Paloma, M. R. Leunda, I. J. Sainz, and F. Fernández. 2001. 'Seroprevalencia de la diarrea viral bovina, Herpesvirus Bovino y Virus Sincicial Respiratorio en Argentina', *Preventive Veterinary Medicine*, 82: 216–20.

OIE. 2018. 'Bovine viral diarrhoea', in *OIE Terrestrial Manual (OIE)*.

———. 2021. 'Infectious bovine Rhinotracheitis', in *Terrestrial Animal Health Code* (2021 © OIE).

———. 2022. 'Minimum requirements for the production and quality control of vaccines', in *OIE Terrestrial Manual*.

Papp, H., B. Laszlo, F. Jakab, B. Ganesh, S. De Grazia, J. Matthijnssens, M. Ciarlet, V. Martella, and K. Banyai. 2013. 'Review of group A rotavirus strains reported in swine and cattle', *Vet Microbiol*, 165: 190–9.

Parreño, V., C. Bejar, A. Vagnozzi, M. Barrandeguy, V. Costantini, M. I. Craig, L. Yuan, D. Hodgins, L. Saif, and F. Fernandez. 2004. 'Modulation by colostrum-acquired maternal antibodies of systemic and mucosal antibody responses to rotavirus in calves experimentally challenged with bovine rotavirus', *Vet Immunol Immunopathol*, 100: 7–24.

Parreño, V., M. V. Lopez, D. Rodriguez, M. M. Vena, M. Izuel, J. Filippi, A. Romera, C. Faverin, R. Bellinzoni, F. Fernandez, and L. Marangunich. 2010. 'Development and statistical validation of a guinea pig model for vaccine potency testing against Infectious Bovine Rhinothracheitis (IBR) virus', *Vaccine*, 28: 2539–49.

Parreño, V., S. A. Romera, L. Makek, D. Rodriguez, D. Malacari, S. Maidana, D. Compaired, G. Combessies, M. M. Vena, L. Garaicoechea, A. Wigdorovitz, L. Marangunich, and F. Fernandez. 2010. 'Validation of an indirect ELISA to detect antibodies against BoHV-1 in bovine and guinea-pig serum samples using ISO/IEC 17025 standards', *J Virol Methods*, 169: 143–53.

Patel, J. R. 2004. 'Evaluation of a quadrivalent inactivated vaccine for the protection of cattle against diseases due to common viral infections', *J S Afr Vet Assoc*, 75: 137–46.

Pecora, A., and M. S. Perez Aguirreburuald. 2017. *Actualización en diarrea viral bovina, herramientas diagnós-ticas y estrategias de prevención* Ciudad Autonoma de Buenos Aires: Ediciones INTA.

Pecora, A., M. S. Aguirreburualde, A. Aguirreburualde, M. R. Leunda, A. Odeon, S. Chiavenna, D. Bochoeyer, M. Spitteler, J. L. Filippi, M. J. Dus Santos, S. M. Levy, and A. Wigdorovitz. 2012. 'Safety and efficacy of an E2 glycoprotein subunit vaccine produced in mammalian cells to prevent experimental infection with bovine viral diarrhoea virus in cattle', *Vet Res Commun*, 36: 157–64.

Pecora, A., M. S. Aguirreburualde, A. Ostachuk, A. Aguirreburualde, M. R. Leunda, A. Odeon, S. Chiavenna, D. Bochoeyer, M. Spitteler, J. L. Filippi, M. J. Santos, S. M. Levy, and A. Wigdorovitz. 2016. 'Erratum to: Safety and efficacy of an E2 glycoprotein subunit vaccine produced in mammalian cells to prevent experimental infection with bovine viral diarrhoea virus in cattle', *Vet Res Commun*, 40: 149.

Pecora, A., D. A. Malacari, M. S. Perez Aguirreburualde, D. Bellido, M. C. Nunez, M. J. Dus Santos, J. M. Escribano, and A. Wigdorovitz. 2015. 'Development of an APC-targeted multivalent E2-based vaccine against Bovine Viral Diarrhea Virus types 1 and 2', *Vaccine*, 33: 5163–71.

Pecora, A., D. A. Malacari, J. F. Ridpath, M. S. Perez Aguirreburualde, G. Combessies, A. C. Odeon, S. A. Romera, M. D. Golemba, and A. Wigdorovitz. 2014. 'First finding of genetic and antigenic diversity in 1b-BVDV isolates from Argentina', *Res Vet Sci*, 96: 204–12.

Pryseley, A., K. Mintiens, K. Knapen, and Y. Van der Stede. 1999. 'Estimating precision, repeatability and reproducibility from Gaussian and non-Gaussian data: a mixed model approach', *J Appl Stat*, 37:1729–47.

Rawdon, T. G., M. G. Garner, R. L. Sanson, M. A. Stevenson, C. Cook, C. Birch, S. E. Roche, K. A. Patyk, K. N. Forde-Folle, C. Dube, T. Smylie, and Z. D. Yu 2018. 'Evaluating vaccination strategies to control foot-and-mouth disease: a country comparison study', *Epidemiol Infect*, 146: 1138–50.

Reppert, E. J., M. F. Chamorro, L. Robinson, N. Cernicchiaro, J. Wick, R. L. Weaber, and D. M. Haines. 2019. 'Effect of vaccination of pregnant beef heifers on the concentrations of serum IgG and specific antibodies to bovine herpesvirus 1, bovine viral diarrhea virus 1, and bovine viral diarrhea virus 2 in heifers and calves', *Can J Vet Res*, 83: 313–16.

Ridpath, J. 2010. 'The contribution of infections with bovine viral diarrhea viruses to bovine respiratory disease', *Vet Clin North Am Food Anim Pract*, 26: 335–48.

———. 2012. 'Preventive strategy for BVDV infection in North America', *Jpn J Vet Res*, 60(Suppl): S41–9.

Ridpath, J. F., S. Bendfeldt, J. D. Neill, and E. Liebler-Tenorio. 2006. 'Lymphocytopathogenic activity in vitro correlates with high virulence in vivo for BVDV type 2 strains: Criteria for a third biotype of BVDV', *Virus Res*, 118: 62–9.

Ridpath, J. F., G. Lovell, J. D. Neill, T. B. Hairgrove, B. Velayudhan, and R. Mock. 2011. 'Change in predominance of Bovine viral diarrhea virus subgenotypes among samples submitted to a diagnostic laboratory over a 20-year time span', *J Vet Diagn Invest*, 23: 185–93.

Ridpath, J. F., J. D. Neill, S. Vilcek, E. J. Dubovi, and S. Carman. 2006. 'Multiple outbreaks of severe acute BVDV in North America occurring between 1993 and 1995 linked to the same BVDV2 strain', *Vet Microbiol*, 114: 196–204.

Romera, S. A., M. Puntel, V. Quattrocchi, P. Del Medico Zajac, P. Zamorano, J. Blanco Viera, C. Carrillo, S. Chowdhury, M. V. Borca, and A. M. Sadir. 2014. 'Protection induced by a glycoprotein E-deleted bovine herpesvirus type 1 marker strain used either as an inactivated or live attenuated vaccine in cattle', *BMC Vet Res*, 10: 8.

Saif, L. J., and F. M. Fernandez. 1996. 'Group A rotavirus veterinary vaccines', *J Infect Dis*, 174(Suppl 1): S98–106.

Schudel, A., C. Van Gelderen, and J. Pardo. 'Grupo ad hoc Biologicos. Fundacion PROSAIA'.

Schweizer, M., H. Stalder, A. Haslebacher, M. Grisiger, H. Schwermer, and E. Di Labio. 2021. 'Eradication of Bovine Viral Diarrhoea (BVD) in cattle in Switzerland: Lessons taught by the complex biology of the virus', *Front Vet Sci*, 8: 702730.

Snowder, G. D., L. D. Van Vleck, L. V. Cundiff, and G. L. Bennett. 2006. 'Bovine respiratory disease in feedlot cattle: environmental, genetic, and economic factors', *J Anim Sci*, 84: 1999–2008.

Spetter, M. J., E. L. Louge Uriarte, A. E. Verna, M. R. Leunda, S. B. Pereyra, A. C. Odeon, and E. A. Gonzalez Altamiranda. 2021. 'Genomic diversity and phylodynamic of bovine viral diarrhea virus in Argentina', *Infect Genet Evol*, 96: 105089.

System, Alabama Cooperative Extension. 'Vaccinations for the beed cattle herd', https://www.aces.edu/blog/topics/beef/vaccinations-for-the-beef-cattle-herd/.

Taffs, R. E. 2001. 'Potency tests of combination vaccines', *Clin Infect Dis*, 33(Suppl 4): S362–6.

Trefz, F. M., A. Lorch, M. Feist, C. Sauter-Louis, and I. Lorenz. 2012. 'Construction and validation of a decision tree for treating metabolic acidosis in calves with neonatal diarrhea', *BMC Vet Res*, 8: 238.

Vanbelle, S. 2016. 'A new interpretation of the weighted kappa coefficients', *Psychometrika*, 81: 399–410.

Viera, A. J., and J. M. Garrett. 2005. 'Understanding interobserver agreement: the kappa statistic', *Fam Med*, 37: 360–3.

Vilcek, S., P. F. Nettleton, D. J. Paton, and S. Belak. 1997. 'Molecular characterization of ovine pestiviruses', *J Gen Virol*, 78 (Pt 4): 725–35.

Walz, P. H., B. W. Newcomer, K. P. Riddell, D. W. Scruggs, and V. S. Cortese. 2017. 'Virus detection by PCR following vaccination of naive calves with intranasal or injectable multivalent modified-live viral vaccines', *J Vet Diagn Invest*, 29: 628–35.

Wang, S., X. Zhao, K. Sun, H. Bateer, and W. Wang. 2022. 'The genome sequence of Brucella abortus vaccine strain A19 provides insights on its virulence attenuation compared to Brucella abortus strain 9–941', *Gene*, 830: 146521.

Wen, Y. J., X. C. Shi, F. X. Wang, W. Wang, S. Q. Zhang, G. Li, N. Song, L. Z. Chen, S. P. Cheng, and H. Wu. 2012. 'Phylogenetic analysis of the bovine parainfluenza virus type 3 from cattle herds revealing the existence of a genotype A strain in China', *Virus Genes*, 45: 542–7.

Woodland, R. 2012. 'The validation of potency tests: hurdles identified by EMA/CVMP/IWP', *Dev Biol*, 134: 69–73.

Wright, P. F. 1999. 'International harmonization of standards for diagnostic tests and vaccines: role of the Office International des Epizooties (OIE)', *Adv Vet Med*, 41: 669–79.

Zhu, Q., B. Li, and D. Sun. 2022. 'Advances in bovine coronavirus epidemiology', *Viruses*, 14: 1109.

Zhu, Y. M., H. F. Shi, Y. R. Gao, J. Q. Xin, N. H. Liu, W. H. Xiang, X. G. Ren, J. K. Feng, L. P. Zhao, and F. Xue. 2011. 'Isolation and genetic characterization of bovine parainfluenza virus type 3 from cattle in China', *Vet Microbiol*, 149: 446–51.

Index

H

I

K

L

M

N

O

P

R

S

T

V

POCKET GUIDES TO
BIOMEDICAL SCIENCES

Series Editor
Lijuan Yuan

A Guide to AIDS, *by Omar Bagasra and Donald Gene Pace*

Tumors and Cancers: Central and Peripheral Nervous Systems, *by Dongyou Liu*

A Guide to Bioethics, *by Emmanuel A. Kornyo*

Tumors and Cancers: Head – Neck – Heart – Lung – Gut, *by Dongyou Liu*

Tumors and Cancers: Skin – Soft Tissue – Bone – Urogenitals, *by Dongyou Liu*

Tumors and Cancers: Endocrine Glands – Blood – Marrow – Lymph, *by Dongyou Liu*

A Guide to Cancer: Origins and Revelations, *by Melford John*

Pocket Guide to Bacterial Infections, *edited by K. Balamurugan*

A Beginner's Guide to Using Open Access Data, by Saif Aldeen Saleh Airyalat and Shaher Momani

Pocket Guide to Mycological Diagnosis, edited *by Rossana de Aguiar Cordeiro*

Genome Editing Tools: A Brief Overview, *by Reagan Mudziwapasi and Ringisai Chekera*

Vaccine Efficacy Evaluation: The Gnotobiotic Pig Model, *by Lijuan Yuan*

The Guinea Pig Model: An Alternative Method for Vaccine Potency Testing, *by Viviana Parreño*

For more information about this series, please visit: https://www.crcpress.com/Pocket-Guides-to-Biomedical-Sciences/book-series/CRCPOCGUITOB